TWENTIETH CENTURY INTERPRETATIONS
OF
MUCH ADO ABOUT NOTHING

A Collection of Critical Essays

Edited by
WALTER R. DAVIS

Prentice-Hall, Inc. A SPECTRUM BOOK *Englewood Cliffs, N. J.*

For My Parents

Acknowledgments

I wish to thank James E. Robinson, Maynard Mack, and my wife for their help in preparing this volume.

WALTER R. DAVIS

Contents

v

Introduction

by Walter R. Davis

I

The years 1598–1600 when *Much Ado About Nothing* was produced find William Shakespeare in his middle thirties, that period variously viewed, according to the viewer's own age, as the fulness of young manhood or the beginnings of middle age. He had a rich life behind him; it might be said that he had one life, a conventional one at his birthplace Stratford-Upon-Avon, behind him, and was midway in a second much more exciting life. He had married young, at eighteen, had married, moreover, a woman eight years his senior, and by the time he was twenty-one had had three children by her. For reasons we can only guess at, he had left his growing family at Stratford, perhaps as early as 1588, and had come up to London to try his fortunes as a player; we know that by 1592 he had turned to writing, for in that year Robert Greene attacked him as an unlettered actor turned playwright and indicated that all three parts of his *Henry VI* had been performed. In the years 1592–98 personal tragedy had touched him: his only son had died in 1596, and he was soon to bury his father, before his own fortieth birthday. But most of what we know about these years indicates success. He had achieved financial prosperity, having purchased a share in the Lord Chamberlain's Company in 1594; and he had completed the transition from employee to stockholder by purchasing the rank of gentleman with coat of arms for his father in 1596 and a fine house, New Place in Stratford, in 1597.

As a poet he had achieved critical esteem and popularity among the better sort with his widely imitated verse narratives *Venus and Adonis* (1593) and *The Rape of Lucrece* (1594), and it is certain that his plays had been popular with all sorts for half a dozen years. Indeed, 1598 might be regarded as the high-point of his reputation, for in that year Francis Meres praised him in print in such fulsome terms as these: "As Plautus and Seneca are accounted the best for Comedy and Tragedy among the Latines: so Shakespeare among the English is the most excellent in both kinds." The *Much Ado* years find Shakespeare midway in his career as dramatist. Behind him, together with the poems, were the *Henry VI* plays, the early tragedies *Titus Andronicus* and *Romeo and Juliet,* the mature histories *Richard II* and *1* and *2 Henry IV,* and the early comedies. Just over the horizon lay *Hamlet* (c. 1601) and the great tragedies. His current interests lay in completing the *Henry IV* tetralogy and, especially, in experimenting with comedy.

The fact that scholars have classified *Much Ado* variously among the "early" or the "middle" comedies is one index of its unique transitional status; for, largely because of its treatment of evil in society, it forms a bridge between the two halves of Shakespeare's career as comedian. After his early experiments with comedy of many sorts, Shakespeare (influenced to some extent by John Lyly and George Peele) began to put his characteristic mark on the form by emphasizing theme and making setting symbolic of it. In the romantic comedies that surround *Much Ado,* the theme has to do with love in society: usually with love thwarted by an oppressive society and succeeding only after that society has been changed. The transformation of the society is usually accomplished by reference outside the existing establishment, as when the oppressed lovers flee to a relatively free pastoral place where they can meet (in *A Midsummer Night's Dream* and *As You Like It*), or when characters from another place enter to ventilate a rather stale society (in *Twelfth Night* and *The Merchant of Venice*). Of these *The Merchant of Venice* (*Much Ado*'s immediate predecessor) stands apart from the others: its society threatens not only love but life itself, and it does so not merely because it is stale or straitening, but because corruption lurks at its heart. *Much Ado* develops what *The Merchant* introduced: here too threats to happiness are seen not as resulting from the legalistic rules of society but as welling up from a deep-seated evil firmly enmeshed in, if not actually caused by, the nature of society itself. And, by replacing the usual two juxtaposed settings of actual and more or less ideal—Athens and the wood, the court and Arden, Venice and Belmont—by the single inescapable world of Messina, it forces attention squarely upon the ineradicable problems of human society, the actual untinged by the possible. In this respect it is one of the most "realistic" of the comedies.

This realistic quality of *Much Ado* will lead forward to the "dark" or "problem" comedies of the succeeding years, plays like *Troilus and Cressida, All's Well That Ends Well,* and *Measure for Measure,* all of which explore the evil inherent in society and therefore tend to share the world, if not the emphasis, of *Hamlet* and the other great tragedies. In the tragedies the envy of Don John will develop into the pure motiveless evil of Iago, just as the staged "rebirth" of Hero will develop into the great mythic rebirth of Hermione that ends *A Winter's Tale*. *Much Ado* and *The Merchant,* are, then, of considerable importance for Shakespeare's total development in that they introduce into his comedy the theme of evil which, after a thorough exploration in the tragedies, will be absorbed back into the comic universe in those final great myths of good and evil, *A Winter's Tale* and *The Tempest*.

II

Although doubts have been raised about the tonal unity of *Much Ado,* the play is, from a technical point of view, more unified than many of the other comedies. For instance, as we have just seen, instead of the two juxtaposed worlds of *A Midsummer Night's Dream, The Merchant of Venice,* and *As You Like It,* we have in *Much Ado* a single setting, Messina, where evil arises and is quelled. Similarly, its clowns Dogberry and the Watch are not detached from the intrigue as Bottom, Lancelot Gobbo, Touchstone, and Feste are, but actually bring about its dénouement by their fumbling discovery of Don John's plot. It is, moreover, demonstrable that its setting, characters, action, imagery, and language all go together with remarkable neatness to create a single clear focus on its main themes. The most striking of these elements is language, which this play—with the self-conscious verbal wit of Beatrice and Benedick and the malapropisms of Dogberry —keeps uppermost in its audience's attention perhaps more than any other of the comedies.

The opening scene of the play immediately focuses on language and lays open some of its implications. It is one of the many ceremonious scenes we shall encounter in this play, Leonato's reception first of the Messenger and then of the victorious army of Don Pedro with Claudio, Benedick, and the defeated bastard brother Don John; and it is fittingly couched in the kind of stiff formal style best illustrated by the Messenger's speech with its Euphuistic balance, alliteration, and word-play:

> He hath borne himself beyond the promise of his age, doing in the figure of a lamb the feats of a lion. He hath indeed better bettered expectation than you must expect of me to tell you how.

The first speeches of Beatrice (who stands on the side-lines) both mock such formality and exhibit in themselves the pliability of language for any purpose by deliberately twisting the Messenger's words, by taking "good service" in the wars as waiting on table, and by perverting "a man, stuffed with all honorable virtues" to "a stuffed man." Similarly, a few moments later after the heroes have entered, Benedick will mock Leonato's circumlocution, that his wife many times told him that Hero is his daughter, by asking, "Were you in doubt, sir, that you asked her?" In this way, the scene establishes two quite different uses of words in addition to the plain conveyance of information or emotion. In one usage, words become a set of conventional counters for use in social intercourse; in the other, a set of toys to be played with. These two functions of language frequently distinguish the two sets of lovers in *Much Ado,* Hero and Claudio on the one hand and Beatrice and Benedick on the other.

Claudio shows himself from the beginning to be a very circumspect young man. So painfully aware is he of the pressures exerted on life by the social amenities that he is frequently unsure whether a person is speaking his true thoughts or merely saying something to fetch him in (I, i, 225–26); and he has accommodated his proceedings so fully to his view of life that it is difficult for others, like Benedick, to discover whether he speaks "in sport" or "truly" (179–80), whether he is requesting a real answer to a question or issuing a call for verbal games (167–70). Both Benedick and Don Pedro show combined amusement and irritation at his slow twisting of "so fine a story" (313), and as he talks so he acts, treading softly: first trying to get Benedick to confirm his own high opinion of Hero's beauty before revealing his love, then ascertaining the lady's financial expectations before requesting Don Pedro to woo her for him. Claudio is an example of the purely social man: polite, little more than a polished surface himself, he shows tender concern for the appearances; therefore he conceives of love not as a grand passion (like Romeo and Juliet) or even as a wayward fancy (like the lovers of *A Midsummer Night's Dream*) but as a social arrangement linked with liking. As such, it must be approached with tact: hence his potentially disastrous decision to have Don Pedro act for him.

Beatrice and Benedick are the satirists of Hero and Claudio. They deliberately mock all conventionalities of speech, disrupt ceremonial occasions with verbal high-jinks, and profess themselves as much enemies of love as Hero and Claudio are its dupes: in Beatrice's words, "I had rather hear my dog bark at a crow than a man swear he loves me" (132–33). They conceive of the relation between the sexes not as a solemn social arrangement but as play, a "merry war" which is the verbal equivalent of sticking out tongues and pulling pigtails. Yet, of course, they are fully involved in what they satirize; they admire each other, and both use wit as a shield against too easy a capitulation of their warm hearts to love. Moreover, their merry war is a conscious pose that they themselves and others come to expect of them, a kind of reverse convention—a fact Benedick shows full awareness of when he asks Claudio,

> Do you question me, as an honest man should do, for my simple true judgment? Or would you have me speak after my custom, as being a professed tyrant to their sex? (167–70)

Act i, scene i presents Messina as a totally conventional world, centering in ceremony, concerned with marriage as a socio-economic compact, and viewing life itself as a matter of keeping up the correct surface appearances. The people here use language as a mask. They delight in deception—whether it be merely the polite indirection necessary to cover real feelings and exist in society tactfully; or a way of innocently amusing one's self; or, finally, a means to evil, as it is for the silent figure in black, Don John (whose very separateness and refusal to indulge in "many words" remind us that the solutions to society's confusions do not lie in withdrawal). The appearances that are so necessary to the social solidarity of Messina are presented in the opening scene as the basis for a series of mistakes that will disrupt it. Not only does society protect evil: deeply imbedded in its nature lie the possibilities for evil.

This fact we see speedily set forth in the two scenes that follow. In scene ii, Antonio has been taken in by appearances so far as to convince himself and Leonato that Don Pedro will woo Hero for himself; in scene iii, Don John decides to use that mistake to disrupt the trust between Claudio and Don Pedro. The crux of this little plot fittingly takes place in the masked ball of ii, i, a ceremony deliberately devoted to an enjoyable hiding of the truth where each of the characters can, as it were, fulfill his desire for hidden action by operating from beneath a mask. Couple by couple, turn by turn, they exhibit the various relations between appearance and reality: Hero sees beneath Don Pedro's mask but not through his manner, for she believes (instructed as she has been by her uncle) that he woos her for himself; Balthasar and Margaret indulge themselves in a love-game like that of Benedick and Beatrice; Ursula forces by flattery the admission of Antonio's identity that she could not achieve by bantering satire (for truth is sometimes only what we are willing to call true); and Benedick, seeking to gall Beatrice by means of his mask, is himself wounded most tellingly. After this sequence of self-generated confusions is ended, Don John confronts Claudio (who pretends to be Benedick) with the supposed fact of Don Pedro's treachery. Belief in such treachery is a plausible mistake in this world, as we see when Benedick himself believes it; but it is just as easily dispelled as it was easily imagined when Don Pedro explains his actions and Claudio's desired match is obtained.

III

The play so far fulfills its title admirably by showing the action of an entire act and one-half coming, precisely, to "nothing," so that it must, as it were, begin all over again. At the same time, it constitutes a remarkably full analogue or pre-image of the main action to come, showing, by a neat and believable progression, how much can be made of nothing; how normal indirection can lead to mistakes; how the villainous can pervert such mistakes, so that they end in confusions of purpose and even of identities; and how, finally, all the hazards may at last dissolve in fulfillment.

The play's main action, which begins to take shape at the end of II, i and II, ii, works by contraries in the attempt of Don John to destroy the love of Hero and Claudio, which is unconsciously countered by the attempt of Don Pedro and his cohorts to create love between Beatrice and Benedick. Both proceed, in a fashion we now see is typical of life in Messina, by deliberately deceptive shows with the intended victim as audience. Claudio sees Margaret dressed as Hero receive Borachio at her window, and Beatrice and Benedick each hear others talk of their great love for each other. The false appearances society fosters are, it seems, ambivalent: they can lead to evil by awakening jealousy, or they can lead to love by arousing the dormant imagination.

The happy deceptions occur first. The gulling of Benedick in II, iii is a carefully contrived reversal-scene; thus it is framed by the contrasting soliloquies of Benedick at the beginning, where he mocks Claudio's love, and near the end, where he ironically applies many of the same themes to himself. Benedick is brought from satirist to lover by a clear dialectical progression, first implicitly by the song (which he, as professed enemy of love and music, rejects) [1] then by explicit statements which he (as, he feels, a man of reason) must accept. The song sets the theme: as itself a representation of changing emotions—"Sigh no more. . . . Converting all your sounds of woe / Into Hey nonny, nonny"—it becomes the first means to change emotions; as a song of consolation to ladies bound to hard-hearted and fickle men, it sets up the arguments by which Benedick will be converted. Following its lead, Don

[1] On this song, see Auden's essay in this volume and Peter Phialas, *Shakespeare's Romantic Comedies* (Chapel Hill: University of North Carolina Press, 1966), pp. 192–93.

Pedro and his co-conspirators first stress their pity for the supposedly desperate love of Beatrice for Benedick and then lament Benedick's hardheartedness, appealing at the same time to his egotism with bantering allusions to his good parts. Benedick is thus presented with two versions of his character, that of the hardhearted satirist he forces himself to be and that of the gentle-hearted lover he could be, and his self-esteem perforce leads him to mold himself into the latter. By doing so, he falls into the trap; in his own words, quite mistakenly, "I should think this a gull but that the white-bearded fellow speaks it." From this point on he becomes so wound up in an ambiguous universe that Beatrice's remarks seem to him to mean the direct opposite of what they convey:

> Ha! "Against my will I am sent to bid you come in to dinner." There's a double meaning in that. "I took no more pains for those thanks than you took pains to thank me." That's as much as to say, "Any pains that I take for you is as easy as thanks." If I do not take pity of her, I am a villain. (266–72)

Act III, scene i, the gulling of Beatrice, is a carefully contrived complement for II, iii: in verse instead of prose, it presents love as a tender and romantic emotion instead of a failure of resistance; and Beatrice's resolve to love, expressed in part of a sonnet at the end, stresses her warm and serious acceptance of affection—and by doing so makes the love affair seem right as well as funny. These paired scenes bring to bear the imagery of music as symbolic of love which James J. Wey explores, and with it the related imagery of dance and birds (in contrast to tempest and discord) which G. Wilson Knight has noticed.[2] But we should add that the bird imagery here usually emphasizes hunting and trapping birds (see, e.g., II, i, 208, 231; iii, 95, 113; and III, i, 24 and 36), for the actions—like music and love in this instance—are traps for the unwary.

The union of the well-deluded pair is suspended for a time while we watch the growing threat to Hero's and Claudio's betrothal. Act III, scenes ii and iv, paired scenes like the earlier two, exhibit first the merry lovers in love and then the movement of the sentimental lovers toward tragedy. Both Benedick and Beatrice now start to experience the pain of alteration—a pain that the former shows by feigning a toothache (to cloak the fact that he has shaved, washed, perfumed, and re-clothed himself), the latter by complaining of a queasy stomach. At this point, too, the imagery of clothing that is to carry much of the

[2] See Wey's essay in this volume, pp. 80–87, and G. Wilson Knight, *The Shakespearean Tempest* (London: Oxford University Press, 1932), pp. 87–93.

theme of appearance and reality in the play comes to the fore.[3] It is from the first associated with love and its changes, in Hero's fussing with her maids over her bridal gown, and in Benedick's gallimaufry of fashions Dutch, French, German, and Spanish—the display of a "fancy that he hath to strange disguises," which Don Pedro is quick to relate to the other variegated "fancy" of love. The comedy of these two scenes is darkened by irony, since the supposedly knowing plotters who mock the witty lovers' entrapment in the web of love "deformed" by "fashion" (as scene iii puts it) do so as their own world is crumbling off-stage. Scene ii ends with Don John's offer to prove Hero's supposed infidelity to Claudio (typically, he does so in order to make his feigned love to Claudio "appear" plain); and scene iv opens the next morning with Hero's preparation for the church-scene that is to prove so disastrous to her.

The broken ceremony at the church in iv, i has been justly celebrated as an instance of Shakespeare's skill in bringing many diverse strands of action into unity, and as such it has been frequently commented on by critics. Walter N. King finds that here the stupidity of those who, like Claudio and Leonato, adhere without question to shallow and rigid codes of behavior, is shown up by others' intelligent examination of the facts; Bertrand Evans similarly analyzes the differing levels of awareness on which the characters in the scene operate, and John Russell Brown applies Claudio's imperfect awareness to the question of love. Different degrees of knowledge naturally show themselves in different stances toward the action, and in this scene we have therefore a wide range of tones, the tragic, the melodramatic, the comic, and so forth, as T. W. Craik has demonstrated; most impressive, as Peter Phialas has shown, is the way in which the melodramatic and the comic manage to coexist harmoniously here. For it is here that the two strands of plot that have been largely held in suspense since the end of Act ii reach completion and interact upon each other.

Claudio begins his revelation of Hero's supposed unchastity with characteristic hesitancy and indirection, for, after all, the "truth" he has to reveal is one he cannot easily assimilate to his view of reality. Therefore, when he finally breaks out in bitterness, what he expresses is a shocked realization that appearances, "seeming," may belie reality. He has loved an image, a concept of Hero rather than the girl herself— hence his dwelling on the "name" of Hero (80–85)—and that image is much more fragile than any woman's actual virtue. Therefore his final response to his new set of "truths" as he stalks out of the church is to retreat completely into the self: "For thee I'll lock up all the gates of

---

[3] See the excerpt from James A. S. McPeek's essay on clothing imagery in this volume, pp. 102–3.

love" (106). Claudio's conduct here is totally narcissistic and self-righteous; at the same time the complete destruction in his mind of a valuable if naive ideal has its pathos.

On these grounds, Friar Francis decides not to repudiate Claudio or
disprove him, but to educate him by means of his own image-making
tendency:

> For it so falls out,
> That what we have we prize not to the worth
> Whiles we enjoy it; but being lacked and lost,
> Why, then we rack the value, then we find
> The virtue that possession would not show us
> Whiles it was ours. So will it fare with Claudio.
> When he shall hear she died upon his words,
> The idea of her life shall sweetly creep
> Into his study of imagination;
> And every lovely organ of her life
> Shall come appareled in more precious habit,
> More moving-delicate and full of life,
> Into the eye and prospect of his soul
> Than when she lived indeed. Then shall he mourn.

As this love-affair was broken by deception, it will again be cemented
by the counterdeception of Hero's death.

The wisdom of the friar's plan is immediately attested. In the latter
part of the scene, we witness the success of just such a play on the imagination. Claudio, who has shown his dark side in cold self-centered-
ness, is replaced by the antiromantics who, though also well-deceived,
now show themselves to be both loving and charitable. Their brief interjections in the early part of the scene establish them as the only ones
whose faith in Hero remains unshaken. After the others leave, this
common faith becomes the foundation of their growing love. Their
scene starts with Beatrice's grief for Hero and Benedick's tender regard for it; when his regard spills over into a blurted confession of
love for Beatrice, the tone changes, and love is ventilated by comedy
as Beatrice gradually drags up her own confession. At the climax, the
strengthening of their love in the ruins of Hero's reasserts itself tragically:

> *Benedick.* Come, bid me do anything for thee.
> *Beatrice.* Kill Claudio.

Some of the chill this curt injunction strikes into Benedick's heart is
mitigated in the ensuing lines by comic hyperbole as Beatrice rants and
raves in fine "tragicall" style, while Benedick vainly tries to stop her

with fragmentary interjections. When she does stop, the lovers reach accord. But it is a complex harmony: by saving the love-confession till now, Shakespeare has placed it in an ironic light. And therefore when Benedick asserts, "Enough, I am engaged," he has merged betrothal with an oath of revenge; the hand offered in pledge of love has become the hand bearing death to his friend as well.

IV

Much of the melodrama—leavened and unleavened—that overtakes the two love-plots in this scene is lightened by a third plot that will bring both to satisfactory conclusions. For the church scene is firmly surrounded by the invincible folly of Dogberry and his cohorts; and, in fact, the whole progress of both loves has been punctuated by his pure farce—in III, iii, III, v, and IV, ii. As Beatrice and Benedick are the manipulators of language, so Dogberry is its abject victim. In his obsessive quest for polysyllables as symbols of status, damnation becomes "salvation," treachery "allegiance," apprehend "comprehend," and sensible "senseless." Intoxicating words are always clouding his mind, as we see when, in III, v, he keeps preventing the plain-speaking Verges from divulging information by insistently re-directing attention to its proper object, his own wisdom:

Verges. Marry, sir, our watch tonight, excepting your Worship's presence, ha' ta'en a couple of as arrant knaves as any in Messina.

Dogberry. A good old man, sir—he will be talking. As they say, "When the age is in, the wit is out." God help us! It is a world to see. Well said, i' faith, neighbor Verges. Well, God's a good man. An two men ride of a horse, one must ride behind. An honest soul, i' faith, sir, by my troth he is, as ever broke bread. But God is to be worshiped, all men are not alike, alas, good neighbor!

Yet it is given to Dogberry's Watch to "comprehend" the villains who threaten all the lovers. As they sit under the eaves (for, in their world, duty and relaxation are as much one as information and self-display, or truth and falsity) they have the whole plot revealed to them by the drunken Borachio. They may be confused as to whether the actual crime committed is "lechery" (or treachery), *lèse-majesté* of Don John, or some form of robbery involving the thief "Deformed"—but they do have the common sense so many of their betters lack to smell a rat, and arrest the villains. The rest of their action, after this initial discovery, consists of fumbling around in an effort to get at the truth and to communicate their findings—just the sort of thing the characters in the main plot are also trying to do in Act IV. In IV, ii, in direct contrast to the preceding church scene, which it parodies in its rhythm of accusation and reply, they manage to make everything come out true, without quite knowing it, when they "Write down Prince John a villain," and when Dogberry, almost speaking for Claudio as well as for himself, cries, "remember that I am an ass, though it be not written down, yet forget not that I am an ass."

After this, it comes as no surprise to find that the malapropisms of Dogberry and his fellows convey the truth in spite of themselves: they have in very fact "comprehended two aspicious persons" (III, v, 50) whose revelations will resolve the plot, and brought order again in the "whole dissembly" (IV, ii). Moreover, it falls to them to make the central comment on the clothing imagery that so dominates the second half of the play; the Watch, picking up Borachio's remark about "what a deformed thief this fashion is," exclaims: "I know that Deformed, a' has been a vile thief this seven year, a' goes up and down like a gentleman" (III, iii, 134–35). Their speech is clumsy but they point unerringly at the central fault in a society where social appearances are taken for reality. Borachio sums up their function when he tells all the other characters on stage: "What your wisdoms could not discover, these shallow fools have brought to light" (v, i, 238–40).

In the first scene of Act v, all the levels of plot—three of them by now—come together and interact. A dramatic reflection of the welcoming scene that opened the play, it recalls to our eyes and ears earlier relationships in order to show how they have changed. For the broken ceremony at the church so disturbed the ceremonious harmony of Messina as to cause several broken relationships; some of the old appearances have worn away. Leonato's opening diatribe to Antonio on the vanity of covering one's real feelings by calm appearances openly contradicts the basis of his and others' actions at the opening of the play (for example, I, i, 18–29, where tears are made by fancy rhetoric to seem signs of joy). And where the earlier occasion led to fulsome welcome, this one proceeds through chilly "Good day's" to a challenge, a challenge that Antonio loads with satire of that empty thief, Fashion, which Claudio and Don Pedro still follow:

> Scambling, outfacing, fashion-monging boys
> That lie, and cog, and flout, deprave and slander,
> Go antiquely and show outward hideousness,
> And speak off half a dozen dangerous words,
> How they might hurt their enemies if they durst—
> And this is all.   (v. i. 94–99)

The hollowness of the old appearances is made stridently evident when Benedick enters and his erstwhile boon companions attempt to re-engage him in the old play of wit and the newer jokes about Beatrice's love for him. These jests are out of fashion, for we (with our higher level of awareness) see here a Benedick, changed by his love and his oath, who stands by in grim silence while Claudio and Don Pedro attempt to envelop all in a cloud of hilarity, then delivers his curt challenge and stalks out.

It is fitting that the planners of Hero's staged disgrace in church should become the objects of the revelations here. It is brought home to them first of all that Leonato believes them to be villains; then that Benedick is in earnest; finally—after skirmishes with Dogberry's web of words—that Hero's infidelity was itself staged by Don John. Here, further, their own particular folly is revealed as well as the general folly of trusting to appearances; for it is just after Don Pedro has said of Dogberry, "This learned constable is too cunning to be understood," that Borachio says the same of him: "I have deceived even your very eyes. What your wisdoms could not discover, these shallow fools have brought to light."

V

The theme of appearance and reality is almost infinitely extendable in Shakespeare's work, depending on how concretely or abstractly one wishes to take it. It can take the form of a social question, as Crick and King take it: how is one to discover any truth or true value for his way of living in a society devoted to the appearances, to living by codes that hide the real? Or that of a psychological question, as Brown takes it: how is one to find the true nature of love in himself beneath the web of appearances and shallow modes of "love"? It can pre-eminently become an epistemological question of knowing true from false, as Hockey takes it to be. At the outer limit it can even be a metaphysical question, as Goddard and Horowitz take it: in an unsubstantial world of empty appearances, how is one to create a reality for himself?

The level of application most firmly supported by the text of the play, with its central concerns of courtship, slander, and law, is the social. Messina is above all a self-absorbed society in which people's concern for appearances argues a self-assurance and smugness that blocks any real sense of connection with the world outside it. The flaw of egotism in Claudio leads him into a snobbish worship of his own idea of a woman and therefore tempts him to misapprehensions and self-righteous accusations. Benedick and Beatrice likewise insulate their egos by nonchalant attitudes, and pay all too much attention to the effects their appearances make on audiences. Dogberry, whose self-importance blinds him to the truth, finds himself an ass because he deemed himself wise. So are they all, asses because they deemed themselves wise; the necessary result of self-absorption is folly.[4]

The theme of appearance and reality is carried over into action in the play by dramatic emphasis on ceremony in public scenes like weddings and ritual welcomes. In this regard, it is worthy of note that all the scenes of Act v reflect—by their settings, their casts, their rhythms, and their actions—earlier scenes of the play in order to show changed relations and, especially, to show by contrast the new sense of reality that has entered the play with Dogberry's revelations. Thus, as v, i reflected i, i in order to show an entirely new and serious Benedick, so v, ii brings Benedick and Beatrice together in the garden where each had been separately deceived into love (ii, iii and iii, i), so that they can tenderly and realistically discuss their achieved affection and, in the process of doing so, satirize the use of "festival terms" to cover up the real. The tiny lyric scene, v, iii, is a ritual corrective to the church scene of iv, i: within its short compass we see night changed to day, the memorial of a funeral to the preparations for a wedding, death to life (lines 5–6), and—after the meaningfully ritual change of Claudio's clothes—atonement for past sins of self-absorption completed and a fresh turn taken toward the future.

Finally, v, iv is, literally and figuratively, the scene where those who had assumed masks in ii, i remove them. Hero's unmasking is presented (in keeping with the preceding ritual scene) as a rebirth of a true and unstained Hero from the ashes of the falsely shamed Hero (lines 59–70). After a new betrothal scene with the right answers, Beatrice unmasks for a rather more complicated discovery: in their patterned responses to each other, Benedick and Beatrice penetrate the illusion woven about them by the others, but use this new truth in their old style to mask the further truth of their love. At this point the "reborn" lovers must step in and force the truth upon them by revealing them as "writ down" asses for love, the reality of their "own hands," as Benedick puts it, revealing their hearts. Once reality has been forced upon them, as upon the others, their shells can break and their self-entrapped affections can flow out to each other. It is fitting that their final pledge to each other combines tenderness and wit, and just as fitting that Ben-

---

[4] For this paragraph—as well as for much else in this Introduction—I am heavily indebted to Maynard Mack's lectures on the play in his undergraduate Shakespeare course at Yale University in 1953–54.

edick finally stops the "merry war" by converting lips from utterers of ambivalent words to plain instruments of love: "Peace! I will stop your mouth. [Kissing her]."

Benedick stops Beatrice's fast chatter by a kiss, offers a generalization on the action—"man is a giddy thing, and this is my conclusion"—and then, calls for the final ceremonial music. That it is Benedick who stops the action and takes over as master of ceremonies widens the significance of the imagery of rebirth with which the last scenes of *Much Ado* abound. For we see that the exchange of appearances for reality retires Don Pedro (representative of the former social establishment, with its firm allegiance to conventions that foster self-absorption) and brings to the fore one who will represent a new outlook, with wit, relative realism, and warmth. In the warm light of this "reborn" society, where music now expresses true joy, comic justice is meted out to all: the intenders of deceit will be punished, while the merely deceived are forgiven; for, as Leonato had earlier replied to Friar Francis' assertion that Hero is innocent,

> So are the Prince and Claudio, who accused her
> Upon the error that you heard debated.   (2–3)

The reader who feels that such easy pardon lets the priggish Claudio off too easily—especially after his disgraceful conduct toward Leonato in v, i—must remind himself that Shakespearean comedy celebrates not so much the triumph of good over evil as that of reality, with all its unidealistic concessions to practicality, over illusion.

Our discussion of *Much Ado* has highlighted a unity that resides chiefly in the meeting of the several elements of drama in the theme of appearance and reality. The setting, restricted to a single circumscribed society whose love for the conventions is expresed in its characteristic ceremonial scenes of wedding, welcome, dance, and so forth, is the realm of appearances. The main characters (even Dogberry) manipulate the appearances to bolster their own self-absorption until forced to face reality. Almost every deed in the play is a deception, for either bad or good ends; hence the action consists mainly of the attempts of three separate groups to pierce through a web of illusions to the truth. The dominant imagery of clothing expresses the cloaking of truth beautifully, and is, moreover, tied both to theme and, especially, to the characteristic actions of disguise and deceit. Finally, the self-conscious language of the play (and its extension into music) is in all cases an ambivalent tool—serving sometimes to reveal truth, more often to screen it by making words serve as counters in conventional games (Claudio and Hero), by playing with words in witty games

(Benedick and Beatrice), or by totally perverting their meanings for everybody's bemusement and amusement (Dogberry and the Watch).

VI

The title page of *Much Ado* as it was first published in Quarto in 1600 assures us that it had been "sundrie times publickly acted by the right honourable, the Lord Chamberlaine his servants," and a slip in the stage-directions in IV, ii tells us that the part of Dogberry was originally taken by the famous clown Will Kempe. The play seems to have been popular until the closing of the theaters in 1642, probably not because of the historical allusions some scholars claim to have found in it,[5] but because of its characters. For, on one of the two occasions of its performance in 1613 during the festivities celebrating the marriage of the Princess Elizabeth it went under the title *Benedicte and Betteris,* and Charles I, who perhaps first saw it acted at his sister's wedding, retitled it "Benedick and Beatrice" in his copy of the 1632 Folio. The merry lovers have, in fact, been responsible for the play's survival on the boards. In 1661, Sir William Davenant lifted the Benedick and Beatrice plot into his adaptation of that darker but similar play, *Measure for Measure,* under the new title *The Law Against Lovers*: Samuel Pepys saw this amalgam on 18 February 1662, and pronounced it "a good play and well performed."

However, we hear nothing of performances of either Davenant's adaptation or the original *Much Ado* for almost eighty years until David Garrick's revival in 1748; Benedick was one of that great actor's favorite parts, and he played it many times during the next few decades opposite the Beatrice of Mrs. Pritchard, who, perhaps inspired by him, is said to have played the part so convincingly that "her uncharacteristic corpulence was always overlooked." [6] From Garrick's time to the present, *Much Ado* has appeared regularly on the boards. Among famous nineteenth-century Benedicks were the great Charles Kemble and George Macready, whose rendering of Benedick in love was said to have been deliciously ludicrous; Helen Faucit was famous for her portrayal of Beatrice, a part she played with regularity from 1846 to 1870. The play's greatest triumph came with Sir Henry Irving's pro-

---

[5] See, for example, G. B. Harrison, *Shakespeare Under Elizabeth* (New York: Henry Holt & Co., 1933), pp. 142–44.

[6] For this and other comments on productions in the eighteenth and nineteenth centuries, see Horace Howard Furness, ed., *A New Variorum Edition of Shakespeare: Much Ado About Nothing* (Philadelphia: J. B. Lippincott Co., 1899), pp. 385–94.

duction of 1882, which ran for two hundred and twelve nights to wild acclaim; Ellen Terry played Beatrice to Irving's Benedick (as she did for the next decade), and was said never to have been more charming —though even her charm could not win over George Bernard Shaw, who, in a later letter to her, called it "Much Adoodle-do. . . . a shocking bad play." [7]

One of the most interesting of the play's many productions in our century was that of the American Shakespeare Festival Theater in Stratford, Connecticut, in the summer of 1957 with Katherine Hepburn and Alfred Drake cast as the witty lovers, new music composed by Virgil Thomson, and a radically changed setting; Sir John Gielgud's New York production of 1959 also aroused considerable controversy in the press.[8]

It is no uncommon paradox that the very same elements that lead to a play's success on the stage—in the case of *Much Ado*, the brilliant wit of Benedick and Beatrice—should highlight its difficulties for literary critics. For the high-jinks of the merry lovers have often served to underline the very uncomic affair of Claudio and Hero which Charles Gildon, the play's earliest critic (1709) found "too shocking for either Tragedy or Comedy";[9] and Hartley Coleridge echoed his judgment over a century later in asserting that "where the ground-work is comic, it is vain to work in flowers of sombre hue." Until comparatively recent times, in fact, criticism of the play has been negative in the main, usually contending either that the merry lovers stole the show from their colorless friends or that, conversely, the Hero-Claudio plot was too distasteful for comedy; even Swinburne, who admired the play extravagantly, found it difficult to forgive "such a pitiful fellow as Count Claudio."

While a few modern critics—notably E. K. Chambers and E. J. West[10]—have continued to find the play unsatisfactory or even unpleasant, many have attempted apologies for the play. Some have concentrated on Claudio, trying to exonerate him (Kerby Neill), or to show that he is not to be taken seriously (Charles Prouty), or that he suffers sufficiently for our forgiveness (Robert G. Hunter). Others have

---

[7] See *Ellen Terry and Bernard Shaw: A Correspondence,* ed. Christopher St. John (New York: G. P. Putnam's Sons, 1931), letter CCLX, 3 June 1903 (p. 293).

[8] For accounts of the Gielgud production, see the reviews by Richard Hayes (*The Commonweal,* LXXI [1959], 104–5) and Gore Vidal (*The Reporter,* XXI, No. 6 [1959], pp. 38–39).

[9] For this and other comments cited in this paragraph, see the *New Variorum* edition, *op. cit.,* pp. 347–62.

[10] See E. K. Chambers, *Shakespeare: A Survey* (London: Sidgwick & Jackson, Ltd., 1925), pp. 134–35 and E. J. West, "Much Ado About an Unpleasant Play," *Shakespeare Association Bulletin* XXII (1947), 30–34.

taken up the other problem and have tried instead to demonstrate the play's intrinsic unity—unity of plot (Wilson), or of tone (Storey, Bradbrook), or, most frequently, of theme (King, Craik, Smith).

But contemporary criticism of *Much Ado* is by no means limited to apologetics. Twentieth-century literary criticism has generally conceived its aim to be recovery, the finding of avenues to older works of literature by relating them to aspects of human experience that endure in any age. Our critics often formalize this aim by focusing on one avenue at a time—on general traits of human character, or myths or rituals that still exist (even if vestigially) in our time, or on archetypes of general human experience hinted at through imagery, or on modern social or economic analogues, or on general philosophical patterns of meaning, and so forth. All these approaches to *Much Ado* will be found in this collection.

That reader will also do well who looks about him thoughtfully at his own society. Confronted by pressures of conformity and status, by the problems of role-playing and of "credibility gaps"; so beleaguered by the manipulation of information in the public media that distinctions between good and bad, valuable and worthless, often seem totally to blur—today's reader is exceptionally well-fitted to grasp the relevance of a comedy about the difficulties of piercing through a web of social roles and delusive words in order to get at the truth.

PART ONE

Interpretations

The Success of *Much Ado About Nothing*

by Graham Storey

May I confess that I only added the first words of my title when I was well into preparing this lecture? Do not mistake me: the riches of the play—the sheer exhilaration of the encounters between Benedick and Beatrice and their arabesques of wit; the superb stupidity of Dogberry and Verges and *their* arabesques of misunderstanding; the skilful weaving and disentanglement of the comic imbroglio—all these are a joy to see and hear, and belong to Shakespeare's most assured writing. But it is a commonplace of criticism that a successful play, like any other work of art, must be a unity: what Coleridge called the Imagination's "esemplastic power" must shape into one its individual forces and beauties. Whether *Much Ado* has this unity was the question that worried me.

It did not worry Shakespeare's contemporaries. The play offered an exciting Italianate melodrama, enlivened by two variegated sets of "humours": the wit-combats and properly-rewarded over-reachings of Benedick and Beatrice, and the low-life comedy of Dogberry and Verges; and remember that George Chapman and Ben Jonson had just started a run of fashionable "humour" plays. As in all proper comedies, the story came out all right in the end. "Strike up, pipers! *Dance.*" The formula ends that other comedy with a similarly riddling title, *As You Like It*; and whatever the differences of tone, the effect does not vary so much from that of the conclusion of *Twelfth Night,* the third of this group of plays written at the turn of the century:

> A great while ago the world begun,
>   With hey, ho, the wind and the rain;

"*The Success of* Much Ado About Nothing" by Graham Storey. *From* More Talking of Shakespeare [*a collection of lectures delivered at the Shakespeare Memorial Theatre's Summer Schools on Shakespeare between 1954 and 1958*], ed. John Garrett (London: Longmans, Green & Co. Ltd., New York: Theatre Arts Books, 1959), pp. 128–43. Copyright © 1959 by Longmans, Green & Co. Ltd. Reprinted by permission of the publishers.

But that's all one, our play is done,
And we'll strive to please you every day.

The humours were what the contemporary audience remembered the play by *"Benedicte and Betteris,"* say the Lord Treasurer's accounts for 1613: and *Much Ado* was almost certainly meant. "Benedick and Beatrice," wrote Charles I in his second Folio, as a second title to the play—exercising a similar Stuart prerogative in renaming *Twelfth Night* "The Tragedy of Malvolio." The "main plot" is clearly being regarded as a kind of serious relief to the much more absorbing comedy. When, with the Restoration, Shakespeare had to face the formidable canons of the neo-Classic critics, this central plot came in for some hard questioning. The criticism was, as we should expect, formal: the *decorum* was at fault. "The fable is absurd," writes Charles Gildon, in 1710, in an essay[1] often reprinted during the eighteenth century; "the charge against Hero is too shocking for tragedy or comedy, and Claudio's conduct is against the nature of love." He is almost equally concerned that the people of Messina do not act and talk, he says, like natives of a warm country.

But, at the turn of this century, one or two critics began to show a quite new uneasiness about the play. They found, not unity, not the almost unblemished gaiety that they found in *As You Like It* or *Twelfth Night*; but jarring tones, a gratuitous suffering and heartlessness in crisis—the Church Scene—that the rest of the play could not wipe out, and a distressing inconsistency in the characters of Claudio, the Prince and Leonato. The critical approaches were different: but the resultant *uncomfortableness* they generated was much the same. And it has undoubtedly left its mark upon many performances since.

The most frequent cause of uneasiness has been to respond to the play as though the protagonists were psychologically real. It is indeed the most expected response, as the dominant mode of the theatre is still naturalism. But it plays havoc with *Much Ado* as *comedy*. Stopford Brooke, writing in 1913[2] as a Bradleyan, shows what happens. He clearly wants to like the play; yet its very centre, the exposure in church, is, he writes, "a repulsive scene." "In it all the characters will be tried in the fire"; and, as a Victorian clergyman of strong, if sensitive views, he tries them. They emerge—Claudio, Don Pedro and Leonato —shallow, wilful, cruel, inconsistent with what they were before; and the play, its centre contaminated, is virtually handed over to Benedick and Beatrice. That, I am convinced, is not how Shakespeare wrote the

[1] *Remarks on the Plays of Shakespeare:* included in *Shakespeare's Poems,* 1710 (supplementary vol. to Rowe's *Works of Shakespeare*).
[2] *Ten More Plays of Shakespeare,* 1913, p. 21.

play. But the figures of the main plot are bound to appear in this light, if we see them as fully-rounded characters and subject them to the tests of psychological consistency. I see them as something much nearer "masks": as not quite so far removed from the formalized figures of *Love's Labour's Lost,* where most of the play's life resides in the plot-pattern and the dance of verbal wit, as many critics have suggested. I will return to this suggestion later. Meanwhile, I only want to insist that the opposite approach—that of naturalistic realism—stretches the play much further than a comedy can go, and makes almost impossible demands of the actors for the last two Acts. It can also lead to a quite ludicrous literalism, as where Stopford Brooke, quoting the magnificent, absurd *finale* of Beatrice's outburst against Claudio after the Church Scene—"O God, that I were a man! I would eat his heart in the market-place"—solemnly comments, "Of course, she would not have done it." [3]

Others though, besides the "naturalist" critics, have found *Much Ado* disturbing: and disturbing because they do not discover in it the unity that I have made my main question. Sir Edmund Chambers,[4] writing fifty years ago, was probably the first to note what he called its "clashing of dramatic planes." "Elements," he wrote, "of tragedy, comedy, tragi-comedy, and farce are thrust together"; and the result is not unity, but "an unco-ordinated welter," a dramatic impressionism that sacrifices the whole to the brilliance of individual scenes or passages of dialogue or even individual lines. Other writers have more recently said much the same: the play's elements are incompatible; the plot too harsh for the characters; it is the wrong kind of romantic story to blend with comedy. "This happy play," as "Q." called it in his Introduction to the *New Cambridge Shakespeare,* 1923, seems, in fact, to be in danger of losing its central place in the canon of Shakespeare's comedies (or it would be, if critics were taken too seriously).

I think that all these critics have seriously underrated the *comic* capacity of both Shakespeare and his audience: the capacity to create, and to respond to, varying and often contradictory experiences simultaneously; to create a pattern of human behaviour from their blendings and juxtapositions; and to obtain a keen enjoyment from seeing that pattern equally true at all levels. I will try to apply this claim to *Much Ado.*

"For man is a giddy thing, and this is my conclusion," says Benedick in the last scene; and this is surely the play's "cause" or ruling theme.

[3] Op. cit., p. 27: quoted by T. W. Craik in *Much Ado About Nothing* (*Scrutiny,* October 1953).
[4] Introduction to *Much Ado* (Red Letter Shakespeare, 1904–8). Reprinted in *Shakespeare: A Survey,* 1925.

"Giddy," a favourite Elizabethan word: "light-headed, frivolous, flighty, inconstant," it meant by 1547; "whirling or circling round with bewildering rapidity" (1593); mentally intoxicated, "elated to thoughtlessness" (in Dr. Johnson's *Dictionary*). *Much Ado* has all these meanings in abundance. And Benedick's dictum, placed where it is, followed by the dance (reminiscent perhaps of the *La Ronde*-like Masked Ball of Act II), suggests eternal recurrence: "Man is a giddy thing"—and ever more will be so. The impetus to two of the play's three plots is the impetus to all the comedies, the propensity to love-making: one plot begins and ends with it; the other ends with it. And the impetus to the third plot, the antics of the Watch ("the vulgar humours of the play," said Gildon,[5] "are remarkably varied and distinguished"), is self-love: the innocent, thoughtless, outrageous love of Dogberry for himself and his position.

Inconstancy, mental intoxication, elation to thoughtlessness: the accompaniment of all these states is deception, self-deception, miscomprehension. And deception, the prelude to "giddiness," operates at every level of *Much Ado*. It is the common denominator of the three plots, and its mechanisms—eavesdroppings, mistakes of identity, disguises and maskings, exploited hearsay—are the major stuff of the play.

In the main plot—the Italian melodrama that Shakespeare took from Matteo Bandello, Bishop of Agen—the deception-theme is, of course, the most harshly obvious. Don John's instrument, Borachio, deceives "even the very eyes" of Claudio and the Prince; Claudio, the Prince and Leonato are all convinced that Hero has deceived them; Hero is violently deceived in her expectations of marriage, stunned by the slander; the Friar's plan to give her out as dead deceives everyone it is meant to.

The deceptions of Benedick and Beatrice in Leonato's garden-bower serve a function as a comic echo of all this. They are also beautifully-managed examples of a favourite Elizabethan device: the over-reacher over-reached, the "enginer hoist with his own petar," the marriage-mocker and husband-scorner taken in by—to us—a transparently obvious trick. (It is a major part of the play's delight that the audience always knows more than the actors: hints are dropped throughout; a Sophoclean comic irony pervades every incident.) Here, the metaphors of stalking and fishing are both deliberately overdone; and the effect is to emphasize that each of these eavesdroppings is a piece of play-acting, a mock-ceremonious game:

> *Don Pedro.* Come hither, Leonato: what was it you told me of to-day, that your niece Beatrice was in love with Signior Benedick?

[5] Op. cit.

Claudio. O! ay: (Stalk on, stalk on; the fowl sits.) I did never think that
lady would have loved any man.[6]

And in the next scene:

> *Ursula.* The pleasant'st angling is to see the fish
> Cut with her golden oars the silver stream,
> And greedily devour the treacherous bait:
> So angle we for Beatrice. . . .
> *Hero.* No, truly, Ursula, she is too disdainful;
> I know her spirits are as coy and wild
> As haggards of the rock.[7]

The contrast between prose and a delicate, artful blank verse makes
sharper the difference of the fantasy each of them is offered. Benedick
is given a superbly ludicrous caricature of a love-sick Beatrice, which
only his own vanity could believe:

Claudio. Then down upon her knees she falls, weeps, sobs, beats her heart,
tears her hair, prays, curses: "O sweet Benedick! God give me patience!
. . ." Hero thinks surely she will die.[8]

And his own response, a mixture of comically solemn resolutions and
illogical reasoning, is equally exaggerated:

I must not seem proud: happy are they that hear their detractions, and
can put them to mending. . . . No; the world must be peopled.[9]

Beatrice has her feminine vanity played on more delicately, but just
as directly: she is given a not-too-exaggerated picture of herself as Lady
Disdain, spiced with the praises of the man she is missing. And her response, in formal verse, clinches the success of the manoeuvre:

> What fire is in mine ears? Can this be true?
> Stand I condemn'd for pride and scorn so much?
> Contempt, farewell! and maiden pride, adieu!
> No glory lives behind the back of such. . . .[10]

"Elated to thoughtlessness" indeed (and particularly after all their earlier wit): but not only by a trick. Benedick and Beatrice are both, of
course, perfect examples of self-deception: about their own natures,
about the vanity their railing hides (and none the less vanity for its

---

[6] II. iii. 98–103.
[7] III. i. 26 ff.
[8] II. iii. 162–5 and 191.
[9] Ibid., 248 ff.
[10] III. i. 107 ff.

charm and wit), about the affection they are capable of—in need of—when the aggression is dropped, about their real relations to each other. This gives the theme of deception in their plot the higher, more permanent status of revelation. Hence much of its delight.

But no one in the play is more mentally intoxicated than Dogberry. He is a king of all he surveys: of Verges, his perfect foil; of the Watch; of the peace of Messina at night. Only words—engines of deception—constantly trip him up; though, like Mrs. Malaprop, he sails on magnificently unaware:

> Dost thou not suspect my place? Dost thou not suspect my years? O that he were here to write me down an ass! . . . I am a wise fellow; and, which is more, an officer; and, which is more, a householder; and, which is more, as pretty a piece of flesh as any in Messina. . . .[11]

With Dogberry, the theme of giddiness, of self-deception, of revelling in the appearances that limitless vanity has made true for him, reaches miraculous proportions.

There is, though, the further meaning of "giddy," also, I suggested, warranted by Benedick's conclusion: "whirling or circling round." The structure of *Much Ado*—the melodramatic Italian love-story, enlivened by two humour-plots of Shakespeare's own invention—follows an established Elizabethan comedy-pattern: Chapman was to use it in *The Gentleman Usher* and *Monsieur d'Olive; Twelfth Night*—allowing for obvious differences in the tone of the central plot—is the obvious successor. Musically, we could call it a theme and variations. But you have merely to consider the Chapman comedies, where the two plots only arbitrarily meet—or Thomas Middleton, who brought in a collaborator to help him with the "echoing sub-plot" of his tragedy, *The Changeling*—to see Shakespeare's extraordinary structural skill here. "Faultless balance, blameless rectitude of design," said Swinburne: he is right, and it was not what most of his contemporaries recognized in *Much Ado*. But it still does not strongly enough suggest the grasp, the intellectual energy, that holds the play together and makes the kind of suggestions about reality in which the Elizabethan audience delighted. Here, again, Benedick's conclusion says more. Not only the play's wit —a microcosm of its total life—whirls and circles, with often deadly effect ("Thou has frighted the word out of his right sense, so forcible is thy wit," cries Benedick to Beatrice in the last Act: it suggests that wit—and wit's author—can destroy or create at will) one of man's main instruments of living; but, in their vibrations and juxtapositions, the three plots do much the same.

[11] IV. 2. 79 ff.

Twice the plots fuse—once to advance the story, once to deepen it—and the achievement gives a peculiar exhilaration. Each time it is something of a shock; and then we see that, within the rules of probability laid down by Aristotle for writers of tragedy (we can validly apply them to comedy too), it is wonderfully right that it should have happened like that.

The first occasion is the discovery by the Watch of the plot against Hero. When they line up to receive their instructions from Dogberry and Verges—on the principle of peace at all costs—it seems incredible that they should ever discover anything. But they do: though, admittedly, Shakespeare has to make Borachio drunk to make it possible. The Watch and their Officers are now locked firmly into the main plot, with all their ripples of absurdity; and the final dénouement is theirs. The innocent saved by the innocent, we may say; or, more likely (and certainly more Elizabethanly), the knaves caught out by the fools. "Is our whole dissembly appeared?" asks Dogberry, as he looks round for the rest of the Court. "Which be the malefactors?" asks the Sexton. "Marry, that am I and my partner," answers Dogberry, with pride.

However we look at it, the impact has clearly changed the status of the villains. "Ducdame, ducdame, ducdame," sings Jaques (it is his own verse) to his banished companions in the Forest of Arden. "What's that *ducdame*?" asks Amiens. "'Tis a Greek invocation to call fools into a circle." Here in *Much Ado,* the knaves have been thrust in with the fools: if it makes the fools feel much more important than they are, it makes the villains much less villainous; or villainous in a way that disturbs us less. This is one device by which the interlocking of plots establishes the play's unity, and, in doing so, creates a new, more inclusive tone.

The entry of the Watch into the centre of the play advances the story. The entry of Benedick and Beatrice, in that short packed dialogue after the Church Scene, where they declare their belief in Hero and their love for each other, seems as though it must do so too; but in fact it does not. Rather, it does not if we see the heart of the play now as Hero's vindication. That is brought about without help from Benedick; and, indeed, Benedick's challenge to Claudio, vehemently undertaken and dramatically presented, is, by the end of the play, treated very casually: only perfunctorily recalled, and easily brushed aside in the general mirth and reconciliation of the ending. Perhaps, then, this scene *removes* the play's centre, puts it squarely in the Benedick and Beatrice plot? That is how many critics have taken it; and what, for example, was in "Q.'s" mind when he wrote of the scene's climax: "'Kill Claudio!' These two words nail the play"; and again, ". . . at this point undoubtedly Shakespeare transfers [the play] from *novella* to drama—to a real spiritual conflict." [12] It is certainly many producers and actors—with understandable temptation—interpreted the scene.

Much Ado demands, of course, a continual switching of interest. We focus it in turn on Benedick and Beatrice, on Hero and Claudio, on Don John and Borachio, on Dogberry and Verges, back to Hero and Claudio, and so on. This gives something of the controlled whirl and circling motion I have commented on. It is also true that this scene between Benedick and Beatrice has a new seriousness; that their shared, intuitive belief in Hero's innocence has deepened their relations with each other, and our attitude towards them. But that is not the same as saying that the play has become something different, or that its centre has shifted. That would seriously jeopardize its design; and, although there *are* flaws in the play, I am sure that its design is what Shakespeare intended it to be.

The play's true centre is in fact neither a plot nor a group of characters, but a theme: Benedick's conclusion about man's giddiness, his irresistible propensity to be taken in by appearances. It is a theme that must embody an *attitude;* and it is the attitude here that provides *Much Ado*'s complexity: its disturbingness (where it does disturb); its ambiguities, where the expected response seems far from certain; but its inclusiveness too, where it is assured. For Shakespeare's approach to this theme at the turn of the century (one could call it the major theme of his whole writing-life, probed at endlessly varying levels) was far from simple. The riddling titles of the group of comedies written within these two years 1598–1600, are deceptive, or at any rate ambiguous. *Much Ado About Nothing, As You Like It, Twelfth Night; or, What You Will*: these can all, as titles, be interpreted lightly, all but cynically, as leaving it to the audience how to take them with a disarming, amused casualness. Or, equally, they can leave room for manoeuvre, include several attitudes, without committing themselves to any. This blending or jostling of sympathies is sufficiently evident in these comedies to have won for itself the status of a convention. Dr. M. C. Bradbrook, who has lovingly pursued all the conventions of the Elizabethan theatre, has called it "polyphonic music";[13] Mr. S. L. Bethell, more directly concerned with the Elizabethan audience, calls their capacity to respond to difficult aspects of the same situation, simultaneously, but in often contradictory ways, "multi-consciousness." [14] *Much Ado* exhibits the one and demands the other in the highest degree.

We must, I think, respond in much the same way as the Elizabethan

[12] Op. cit., pp. xiii and xv.
[13] In *Shakespeare and Elizabethan Poetry*, 1951: the title of Chapter **X**.
[14] In *Shakespeare and the Popular Dramatic Tradition*, 1948, *passim*.

audience did, if we are to appreciate to the full the scene between Benedick and Beatrice in the church; and that oddly-tempered, but still powerful scene of Leonato's outbursts to Antonio at the beginning of Act v. For both these scenes, however different—the first is set in a half-comic key, the second employs a rhetoric that is nearer the formally "tragic"—employ deliberate ambiguities of tone and demand a double response.

I will examine the Benedick and Beatrice scene first. Here, Shakespeare clearly means us to sympathize with Beatrice's vehement attacks on Claudio on Hero's behalf, and with the mounting strength of Benedick's allegiance to her. At the same time, he overdoes the vehemence, exposes it to the comedy of his wry appraisal, brings both characters to the edge of delicate caricature. The scene's climax (I have quoted "Q.'s" remarks on it) has been taken to show the maximum deployment of Shakespeare's sympathy. It also exhibits perfectly his comedy. Benedick and Beatrice have just protested they love each other with all their heart:

> *Benedick.* Come, bid me do anything for thee.
> *Beatrice.* Kill Claudio.
> *Benedick.* Ha! not for the wide world.
> *Beatrice.* You kill me to deny it. Farewell.[15]

Superbly dramatic: these fresh shocks in three lines, and, with each, a new insight into human nature; but also highly ironical. To demand the killing of Claudio, in the world established by the play, is ridiculous. To refuse it at once, after the avowal to do *anything*, equally so, however right Benedick may be ethically (and the irony demands that he refuse *at once:* I am sure Dr. Bradbrook[16] is wrong in saying that he hesitates). And for Beatrice, upon this refusal, to take back her heart, having given it a moment before, completes the picture: passionately generous to her wronged cousin, if we isolate the exchange and treat it as a piece of magnificent impressionism; heroic, absurd and a victim to passion's deception, if we see it—as we surely must—within the context of the whole play.

Mr. T. W. Craik, in an admirably close analysis of *Much Ado* in *Scrutiny*,[17] makes this scene between Benedick and Beatrice a pivot of the play's values. It is, he says, " 'placed' by the scene's beginning [i.e. the earlier events in the church]. Putting the point crudely, it represents the triumph of emotion over reason; the reasonableness of Friar

[15] IV. i. 293–296.
[16] Op. cit., p. 183.
[17] October 1953, op. cit.

Francis's plan for Claudio and Hero. . . ." [18] I agree with him when
he goes on to say that "emotion's triumph" is laughable in Benedick
(though I think he exaggerates its extent). But surely it is an oversim-
plification to identify Shakespeare's attitude—as he seems to do more
explicitly later in his essay[19]—with the Friar's common sense. The
Friar is essential to the plot (and much more competent in guiding it
than his brother of *Romeo and Juliet*); and his calm sanity admirably
"places" Leonato's hysteria in the Church Scene. But the whole spirit
of the play seems to me antagonistic to any *one* attitude's dominating
it. And the second scene I want to examine—Leonato and Antonio in
v. i.—appears to bear this out.

For here Antonio begins as the repository of the Friar's wisdom, as
the Stoic, calming Leonato down. Yet, as experience floods in on him
—the memory of wrong in the shape of Don Pedro and Claudio—he
too becomes "flesh and blood," and ends up by out-doing Leonato:

> What, man! I know them, yea,
> And what they weigh, even to the utmost scruple,
> Scambling, out-facing, fashion-monging boys,
> That lie and cog and flout, deprave and slander,
> Go antickly, show outward hideousness,
> And speak off half a dozen dangerous words,
> How they might hurt their enemies, if they durst;
> And this is all!

> *Leonato.* But, brother Antony,—[20]

The roles are neatly reversed. But the invective is too exuberantly
Shakespearian to be merely—or even mainly—caricature. Can we say
the same of Leonato's outburst that begins the scene?

> I pray thee, cease thy counsel,
> Which falls into mine ears as profitless
> As water in a sieve: Give not me counsel; . . .[21]

Considered realistically, it must make us uneasy. Leonato knows (Antonio does not) that Hero is in fact alive: to that extent, most of his
emotion is counterfeit. Again, we remember his hysterical self-pity of
the Act before, when his attitude to his daughter was very different:

[18] p. 308.
[19] p. 314.
[20] v. i. 92–9.
[21] v. i. 3–5.

> Do not live, Hero; do not ope thine eyes;
> . . . Griev'd I, I had but one?
> Chid I for that at frugal nature's frame?
> O! one too much by thee. Why had I one?
> Why ever wast thou lovely in mine eyes? [22]

To some extent, he is still dramatizing himself in this scene, still enjoying his grief. But his language is no longer grotesque or self-convict-
ing, as that was. He echoes a theme—"experience against auctoritee,"
the Middle Ages called it—which in *Romeo and Juliet* had been
nearer a set piece:

> *Friar Laurence.* Let me dispute with thee of thy estate.
> *Romeo.* Thou canst not speak of that thou dost not feel. . . .[23]

but here it has a new authenticity in movement and image:

> for, brother, men
> Can counsel and speak comfort to that grief
> Which they themselves not feel; but, tasting it,
> Their counsel turns to passion, which before
> Would give preceptial medicine to rage,
> Fetter strong madness in a silken thread,
> Charm ache with air and agony with words.[24]

Again, as with Benedick and Beatrice, the whole scene, ending with
the challenge of Claudio and the Prince to a duel, presents a mixture
of tones: appeal to our sympathy, exaggeration which is on or over the
edge of comedy.

Both these scenes, peripheral to the main plot, but of the essence of
the play's art, demand, if they are to be fully appreciated, a complex
response. What, then, of the crux of *Much Ado,* the shaming of Hero
in church? On any realistic view it must, as has been said, be a repulsive scene: an innocent girl slandered and shamed by her betrothed,
with apparently deliberate calculation, during her marriage-service,
and in front of her father—the city's Governor—and the whole con-
gregation. However we see it, Shakespeare's writing here is sufficiently
powerful to give us some wincing moments. No interpretation can take
away the shock of Claudio's brutal

[22] IV. i. 125, 129-32.
[23] III. iii. 62-3.
[24] V. i. 20-6.

> There, Leonato, take her back again:
> Give not this rotten orange to your friend;[25]

or of the Prince's heartless echo:

> What should I speak?
> I stand dishonour'd, that have gone about
> To link my dear friend to a common stale.

The clipped exchange between Leonato and Don John that follows
seems to give the lie the ring of finality, to make false true in front of
our eyes:

> *Leonato.* Are these things spoken, or do I but dream?
> *Don John.* Sir, they are spoken, and these things are true.

The generalizing assent, helped by the closed-circle form of question
and answer, has a claustrophobic effect on both Hero and us (I think
of the nightmare world of "double-think" closing in in Orwell's *Nineteen Eighty-Four*: this is a verbal nightmare too). Momentarily, we
have left Messina and might well be in the meaner, darker world of
that later play of similarly quibbling title, but much less pleasant implications, *All's Well That Ends Well*. There "these things" are commented on by a Second French Lord, who knows human nature; knows
Parolles and his hollowness: "Is it possible he should know what he is,
and be that he is?"; and Bertram and his meanness: "As we are ourselves, what things are we!" ("Merely our own traitors," adds the First
Lord, almost redundantly.)

Then, with a jolt, we remember that "these things" are *not* true.
They are not true *in* the play, which is the first thing to remind ourselves of, if we wish to preserve the play's balance as comedy. For in
the later and so-called "Problem Comedies" (tragi-comedies, I prefer
to follow A. P. Rossiter in calling them)—*All's Well, Troilus and
Cressida, Measure for Measure*—such accusations *are* true, or would
be true if those accused of them had had their way—had not been
tricked into doing something quite different from what they thought
they were doing (Cressida comes into the first category; Bertram and
Angelo into the second). But here the characters are playing out an
act of deception, each of them (except Don John) unaware in fact of
what the truth is. To that extent, they are all innocent, Claudio and
the Prince as well as Hero: played on by the plot, not (as we sense
wherever tragic feeling enters) playing it, willing it. The *situation* is
in control.

[25] IV. i. 31–2.

Secondly, they are not true *outside* the play. To state that at all probably sounds absurd. But genuine tragic feeling in Shakespeare forces its extra-theatrical truth on us: continuously in the tragedies, spasmodically—but still disturbingly—in the tragi-comedies. We know only too well how permanently true are *Hamlet, Othello, Macbeth*. But the exposures of the tragi-comedies (Hero's shaming by no means exhausts the *genre*) inflict on us truths about human nature—we may prefer to call them half-truths. "But man, proud man," cries out Isabella (and she has every justification),

> Drest in a little brief authority,
> Most ignorant of what he's most assur'd,
> His glassy essence, like an angry ape,
> Plays such fantastic tricks before high heaven
> As make the angels weep; who, with our spleens,
> Would all themselves laugh mortal.[26]

Here, all the possibilities of human nature are on the stage. We are *involved with* the people who are hurt or betrayed or even exposed (Angelo, as he cries out on the "blood" that has betrayed him, is potentially a tragic figure); we are involved too in the language and its searing comments on human frailty or baseness.

But go back to the scene in *Much Ado*, and, after the first shock, we are no longer fully involved. First, because the identities of Hero and Claudio have been kept to an irreducible minimum. That is why I earlier called them "masks." They have a part to play in a situation that is the climax to the whole play's theme; but they have not the core of being—or of dramatic being—which suffers or deliberately causes suffering. It would be quite different—ghastly and impossible—to imagine Beatrice in Hero's position.

And, secondly, the whole scene's deliberate *theatricality* lessens our involvement and distances our emotions. It emphasizes that it is, after all, only a play and intended for our entertainment;[27] we know that the accusation of Hero is false and—as this is a comedy—is bound to be put right by the end. First Claudio, then Leonato, takes the centre of the stage: the effect is to diminish any exclusively tragic concern for Hero, as we appraise the responses of the other two. There can be no doubt about Leonato's: it is highly exaggerated and hovers on the edge of caricature. We recognize the tones from *Romeo and Juliet*. There, vindictive, absurd old Capulet hustles Juliet on to a marriage

---

[26] *Measure for Measure*, II. ii. 117–23.
[27] S. L. Bethell makes the same point about the ill-treatment of Malvolio: op. cit., pp. 33–4.

she abhors; and then, in a stylized, cruelly comic scene, is shown (with his wife and the Nurse) over-lamenting her when she feigns death to avoid it. Shakespeare has little pity for this kind of selfishness. Here, as Leonato inveighs against his daughter—now in a swoon—we have self-pity masking itself as righteous indignation: the repetitions show where his real interest lies:

> But mine, and mine I lov'd, and mine I prais'd,
> And mine that I was proud on, mine so much
> That I myself was to myself not mine,
> Valuing of her . . .[28]

Yet, as he goes on, the tone alters, as so often in this volatile, quick-changing play:

> . . . why, she—O! she is fallen
> Into a pit of ink, that the wide sea
> Hath drops too few to wash her clean again,
> And salt too little which may season give
> To her foul tainted flesh.[29]

That is still over-violent, but the images of Hero's stain and of the sea failing to make her clean introduce a different note. We have heard it in Claudio's accusation:

> Behold! how like a maid she blushes here.
> O! what authority and show of truth
> Can cunning sin cover itself withal. . . .[30]

and in his outburst against seeming: "Out on thee! Seeming! I will write against it. . . ."

Again, there is more here than his earlier, calculated stage-management of the scene. It is as though the situation has suddenly taken charge, become horribly true for a moment; and as if Shakespeare has injected into it some of the disgust at sexual betrayal we know from the dark Sonnets and from the crises of a host of later plays: *Measure for Measure, Troilus and Cressida, Hamlet, Othello, Cymbeline.*

This apparent intrusion of something alien—seemingly personal—into the very centre of the play was what had led me to doubt its success. I was wrong, I think (and it follows that I think other doubters are wrong), for three reasons. First, the intrusion, the cold music,

[28] IV. i. 138–41.
[29] Ibid., 141–5.
[30] IV. i. 34–6.

is only a touch; one of several themes that make up the scene. Its language is harsh, but chimes in with nothing else in the play: no deadly vibrations or echoes are set up. Compare Claudio and Leonato with Troilus or Isabella, or, even more, with Hamlet or Othello, in whose words we feel a wrenching, an almost physical dislocation of set attitudes and beliefs: and the outbursts here have something of the isolated, artificial effect of set speeches.

Secondly, the play's central theme—of deception, miscomprehension, man's "giddiness" at every level—is dominant enough to claim much of our response in *every* scene: including this climax in the church that embodies it most harshly, but most fully. And, in its manysidedness and "many-tonedness," this theme is, as I have tried to show, one well within the tradition of Elizabethan comedy.

Thirdly, and lastly, the *tone* of *Much Ado*—animated, brittle, observant, delighting in the ado men make—does not have to stretch itself much to accommodate the moments of questioning in the church. And this tone is ultimately, I think, what we most remember of the play: what gives it its genuine difference from *As You Like It* and *Twelfth Night*. Although two of its most loved figures are Warwickshire yokels (and nothing could change them), the aura of Bandello's Italian plot pervades the rest. The love of sharp wit and the love of melodrama belong there; so do the sophisticated, unsentimental tone, and the ubiquitous, passed-off classical references: to Cupid and Hercules, Leander and Troilus. Gildon was wrong: in essentials, the people of Messina *do* act and talk like natives of a warm country.

The tone I mean is most apparent—most exhilarating and most exacting—in the wit-flytings between Benedick and Beatrice; but it dominates the word-play throughout: and this is one of the most word-conscious and wittiest of all Shakespeare's comedies. If I have said little about the words and the wit, this is because no one was more at home there, and could better communicate his enjoyment of them, than the late A. P. Rossiter: and you can read his lecture[31] on the play from one of the last of his memorable Shakespeare courses at Cambridge. My own debt to him will be very clear to all of you who heard his many lectures here at Stratford.

[31] One of twelve lectures given at Stratford and Cambridge [published in A. P. Rossiter, *Angel with Horns* (London: Longmans, Green & Co. Ltd., 1961)—Ed.].

Messina

by John Crick

"The fable is absurd," wrote Charles Gildon in 1710, and most of us would agree. Yet there is the effervescent presence of Beatrice and Benedick and the engaging stupidity of Dogberry and Verges to assure us that all is not dross. Coleridge was convinced that this central interest was Shakespeare's own, his motive in writing the play, and the "fable" was merely a means of exhibiting the characters he was interested in. This may have been the attitude of audiences in Shakespeare's time: as early as 1613, the play was referred to as "Benedicte and Betteris." Can we summarise the play in this way: a few good acting parts standing out against the unsatisfactory background of a preposterous Italian romance? I think not.

Most of the play's critics have seized on the apparent absence of any unifying dramatic conception: the play fluctuates uneasily, it is said, between tragedy, romance, and comedy and never establishes a convincing dramatic form for itself. In these circumstances there are too many inconsistencies of plot and character and, in particular, in the presentation of Claudio and Hero: they begin as the hero and heroine of a typical Italianate romance and, under the growing dominance of Beatrice and Benedick in the play, become—rather unconvincingly— the perpetrator and victim respectively of a near-criminal act. Beatrice and Benedick throw the play off its balance.

It is a truism that criticism should be concerned with what a work of art is, and not with what it ought to be. In the case of "Much Ado," however, it is one worth remembering, for preconceptions about form, plot, and character, and the other components of a play, have so often obscured what is unmistakably there, and shows itself in the very first scene of the play: the precise delineation of an aristocratic and metropolitan society. This is done with a thoroughness and depth which is beyond any requirement of a romantic fable in the tradition of Ariosto

"Messina" [Editor's title] by John Crick. From "Much Ado About Nothing," The Use of English, XVII (London: Chatto & Windus, 1965), 223–27. Copyright © 1965 by Chatto & Windus. Reprinted and retitled by permission of the publisher.

and Bandello, and beyond the demands of a plot merely intended to exhibit the characters of Beatrice, Benedick, Dogberry and Verges, in the way that Coleridge suggested.

The opening scene of the play establishes for us the characteristic tone of Messina society. Don John's rebellion has been successfully put down and the victors are returning to Messina with their newly-won honours. It is significant that, in spite of the fact that Don John still exists to cause trouble, there is no serious discussion of the reasons for or consequences of the rebellion. War is regarded as something that might deprive society of some of its leading lights—Leonato asks the messenger "How many gentlemen have you lost in this action?"—and enhance the status of others. The messenger informs us that no gentlemen "of name" have been lost, and Claudio and Benedict have fought valiantly and achieved honour. War is a gentlemanly pursuit, a game of fortune—nothing more.

This first conversation of the play has a studied artificiality which seems to bear out this reading of the situation. The language is sophisticated and over-elaborate, as if it has been cultivated as an end in itself, and not as a vehicle for the discussion of serious matters. Leonato's sententiousness may be that of an old man; yet it fits naturally into the play's elaboration of words:

> "A victory is twice itself when the achiever brings home full numbers."
> "A kind overflow of kindness: there are no faces truer than those that are so washed. How much better is it to weep at joy than to joy at weeping!"

Even the messenger—a person of humble origin, we presume—has caught the infection and uses euphuistic phraseology:

> "He hath borne himself beyond the promise of his age, doing, in the figure of a lamb, the feats of a lion: he hath indeed better bettered expectation than you must expect of me to tell you how."
> "I have already delivered him letters, and there appears much joy in him; even so much that joy could not show itself modest enough without a badge of bitterness."

This initial impression—of ornate language as the normal conversational mode in upper-class Messina society—is confirmed by the rest of the play: there is an abundance of antitheses, alliterations, puns, euphuisms, repetitions and word-patterns. The imagery has a similar artificiality and tends to consist of the prosaic and the conventional, rather than the striking. Prose, rather than verse, is the natural medium for conventional talk and ideas, and it is therefore not surprising that there is far more prose in "Much Ado" than is normal in a Shakespearean comedy.

In such a society, Beatrice and Benedick are naturally regarded as prize assets. They, too, relish talking for effect—although they do it with far more wit and vigour than the others, whose speeches are usually lifeless and insipid. If Don John's rebellion has not been taken seriously, as we suspect, it is probably because the "merry war" between Beatrice and Benedick is of far more interest to a fashionable society which, as such societies do, regards a war between the sexes as a subject of perennial fascination. Beatrice, as Benedict says, "speaks poniards" and "every word stabs"; and yet no harm is done. No Messina gentleman is likely to be deprived of his life by "paper bullets of the brain." Yet, one of the play's ironies is that it leads us to doubt this: considerable damage is done by the mere power of words. (It is another of the play's ironies that Beatrice's "Kill Claudio"—an unusually straightforward command—is motivated by charitable feelings.) Hero—the main victim—comments on this power: "one doth not know How much an ill word may empoison liking."

Conventional people and societies often relish the unconventional as a safety-valve for repressed instincts. In a society such as Messina's, where the instincts for life are in danger of being drained away in small talk, Beatrice and Benedick offer this outlet. Their conventional role is to appear unconventional. Where the normal fashionable marriage is based on economic interests, and is ironically the end-product of romantic notions of love centred on physical appearance, a "partnership" of antagonisms and verbal bombardments will offer a vicarious satisfaction to onlookers. Beatrice and Benedick know this and, like court jesters, give society what it wants, until it has to be jolted out of its complacency when near-tragedy strikes. Shakespeare gives us enough indications that they have this self-awareness. Benedick says, "would you have me speak after my custom, as being a professed tyrant to their sex?" and "The prince's fool! Ha? It may be I go under that title because I am merry." Claudio quite rightly says that Benedick "never could maintain his part but in the force of his will." As for Beatrice, the true basis of her supposed misogamy is indicated by her assertion that "manhood is melted into courtesies, valour into compliment, and men are only turned into tongue, and trim ones too: he is now as valiant as Hercules that only tells a lie and swears it." Her war is against the Count Comfects of this world and men of "earth." Benedick, too, is looking for a superior marriage partner: "till all graces be in one woman, one woman shall not come in my grace."

Against this background are presented the conventional hero and heroine—Claudio and Hero. It is untrue to say that Beatrice and Benedick steal the limelight from them because Claudio and Hero never hold it. Hero is far too nebulous a figure, and Claudio is made unattractive from the start. He is a typical young gentleman of Messina society—"a proper squire," as Don John says—with an ear and eye to fashion. His romantic notions of the opposite sex—"Can the world buy such a jewel?"—are grounded in a realisation of the economic basis of fashionable marriages in Messina society—"Hath Leonato any son, my lord?"—(In Bandello, Leonato is poor). We are reminded of Bassanio's "In Belmont is a lady richly left And she is fair . . ." in *The Merchant of Venice*. The shallowness of Claudio's attitude to life is betrayed by his every action. He leaves the wooing of Hero to Don Pedro, and then abandons the courtship with inordinate haste, taking a mere eleven lines to convince himself of the truth of Don John's allegation against Don Pedro, even though the latter has "bestowed much honour" on him. He is merciless and revengeful when his pride has been wounded by the supposed betrayal, and punishes Hero and her father with sadistic exuberance in the "wedding scene"—"a rotten orange," he calls Hero. He refuses to abandon his normal flippancy when faced by an angry Benedick in the scene where the latter challenges him. Even when he knows he has done wrong, he refuses to admit his full guilt—"yet sinned I not but in mistaking." He is willing to accept another marriage offer without a moment's hesitation, perhaps spurred on by the knowledge that the girl is another heir; and his mourning for Hero is very formal and ritualistic, and couched in artificial terms and rhyming verse which has a false ring. Significantly, whereas Bandello emphasises the hero's repentance, this is made a minor affair in Shakespeare, and I can see no evidence for W. H. Auden's view, expressed in an *Encounter* article, that Claudio "obtains insight into his own shortcomings and becomes, what previously he was not, a fit husband for Hero." Such a character is incapable of development for Shakespeare offers him as a postulate, a representative type.

In Claudio, therefore, the worst aspects of Messina society are revealed: its shallowness, complacency, and inhumanity. There is nothing absurd about Beatrice's "Kill Claudio"; in terms of the situation that has been revealed to us, the reaction is a natural one. Where Messina conventions are fallible—and Beatrice as a woman, in a predominantly masculine ethos of courtship, games and war, is particularly qualified to speak here—is in questions of love, marriage, and the relationship between the sexes. Beneath her raillery, Beatrice shows a realistic and discriminating attitude to the subjects. She won't accept the choice of others for a husband, ironically remarking, "Yes, faith; it is my cousin's duty to make curtsy and say, 'Father, as it please you.' But yet for all that, cousin, let him be a handsome fellow, or else make another curtsy, and say 'Father, as it pleases me' "; she rejects romantic notions of the opposite sex—"Lord, I could not endure a husband with a beard on his face"; and, by implication, she won't accept a business marriage. (Benedick's attitude to marriage is similarly realistic—"the world must be peopled"). Hers is a sane perspective on events, an application of generosity and sympathy in a society dominated by ultimately inhumane standards. Her feminine charity triumphs, as Portia's mercy does in *The Merchant of Venice*. Benedick becomes acceptable to her when he symbolically joins his masculine qualities to her feminine principles by taking up, however reluctantly, her attitude to Claudio, and thus shows himself to be, in her eyes, of a finer "metal" than the average Messina male. Ironically, the plotting which separated Claudio and Hero brings them together, their true feelings breaking through their conventional jesters' roles, and it is Beatrice's clear-sightedness which triumphs over all the pattern of misunderstandings, deceptions, and self-deceptions which make up the play. (This patterned and stylised aspect of the play is very marked in the plot, characterisation, and langauge: consider, for example, the balancing of the two scenes in the church; the characterisation in pairs: the artificiality of the masque and the mourning scene; and the rhetorical devices of most of the language.)

The incapacity of Messina society is also exposed, at another level, by Dogberry and Verges. Dogberry, like his superiors, adopts the mode of language and behaviour he conceives to be fitting to his position. When it comes to a real-life drama, he is as patently useless as Claudio. He displays condescension towards Verges and all the pompousness of authority: "I am a wise fellow, and, which is more, an officer, and, which is more, a householder . . ." Claudio, too, has "every thing handsome about him." Dogberry has caught the Messina infection of pride and self-centredness, that self-centredness which makes Leonato —the perfect host at the beginning of the play—wish Hero dead because of the way in which she has shamed him. (Isn't there something more than just a resemblance of name between him and Leontes and Lear?)

Essentially, the play is, I believe, about the power for evil that exists in people who have become self-regarding by living in a society that is closely-knit and turned in on itself. The corruption is usually that of town and city life. (Significantly, Shakespeare's story does not fluctuate between town and country as Bandello's does.) A moral blindness is generated that, if not evil itself, is capable of evil consequences. The agency of evil in this play is not outside, but within. The ostensible villain of the piece—Don John—is a mere cardboard figure who, excluded from a world of flatteries and courtesies, has resorted to "plain-dealing" villainy. He may be an early sketch for Iago and Edmund but he lacks their intelligence and flair, and Shakespeare has wisely kept him within the narrow bounds appropriate for comedy. The real origin of the crime is not jealousy, sexual or otherwise, but blind, consuming egotism which expresses itself in a studied artificiality, and at times flippancy, of both language and attitude. Later, Shakespeare was to take the same theme and mould it into tragedy. In the world of Othello, Lear, and Gloucester, the consequences of pride and self-centredness are catastrophic. The ultimate is perhaps *King Lear*—another "much ado about nothing"—where Lear, like Claudio, could say "Yet sinned I not but in mistaking."

Imagining the Real

by David Horowitz

Benedick, didst thou note the daughter of Signior Leonato?
I noted her not; but I looked on her.

Note notes, forsooth, and nothing.

(Much Ado About Nothing)

The term for perception in *Much Ado* is "noting." To "note" is to take note of, to note specially, to set down in the memory, to pay serious attention to; a "noting" is a special kind of perception therefore. Such a perception may yield more than is seen by merely looking and, therefore, from the point of view of the satiric disengagement and its relentless realism, less. The term itself suggests this possibility, because the Elizabethans sounded the word "noting" the same way as "nothing" (i.e. both words were *noting*).

Early in the action of *Much Ado*, Don Pedro meets Benedick moments after Claudio has left him. Claudio has been complaining of his supposed betrayal by the prince (a betrayal suggested by experience which has been falsely noted to him). Benedick engages the prince in a discussion of this betrayal and Claudio's reaction to it (in order presumably to clear things up), but the prince misnotes his words and misses his meaning, and the misunderstanding between them goes unpatched for the time being:

> *Benedick.* I told him [Claudio], and I think I told him true, that your Grace
> had got the good will of this young lady. (II. 1)

The prince notes a meaning in these words that Benedick does not intend, namely, that his graciousness had got the goodwill of the young lady for Claudio (for Don Pedro had previously indicated that he would intervene on Claudio's behalf). Benedick, of course, means quite

"Imagining the Real" by David Horowitz. From Shakespeare: An Existential View (New York: Hill and Wang, Inc., 1965), pp. 19-36. Copyright © 1965 by David Horowitz. Reprinted by permission of the publisher.

the opposite. He means that his Grace, i.e. Don Pedro, had won the favour of Hero for himself. Thus, a nothing meaning, a meaning that was not intended, becomes the whole meaning, and obscures the sense of the speech.

It is this kind of error, in fact, that dominates the major actions of the play, especially those leading up to the near tragedy of Claudio and Hero; for the way to this misfortune is paved entirely by "events" which are misnoted notings (nothings). Hence the title of the play with its *double entendre:* Much Ado About Not(h)ing.

The "tragedy" of the play is triggered when Claudio and Don Pedro note an experience, namely, the presence of two figures (Borachio and Margaret) in Hero's chamber, that is not an experience at all, because it is something other than what they take it to be. They take the scene to be evidence that Hero has betrayed Claudio. They do this, because Don John, who has created the scene, notes it to them in a special way. He has created not only the scene (for he has instructed Borachio in what to do) but its significance as well, which he does by naming or noting the characters to the audience (Claudio and Don Pedro): this shadow is Hero. But this shadow is not Hero. Noted differently, the experience is a mere nothing.

Not surprisingly for a play whose central action turns on such an incident, *Much Ado* is persistent in its pursuit of the question of appearances. In the beginning, Claudio notes the appearance of more sweetness and modesty in Hero than does Benedick: "Can the world buy such a jewel?" he says; but only a short time later he is ready to believe that she has been won by Don Pedro and so bids every eye "negotiate for itself," for "beauty is a witch." Later, after he imagines he has "seen" himself betrayed again, he can note in her blushes only "guiltiness, not modesty":

> *Claudio.* You seem to me as Dian in her orb,
>   As chaste as is the bud ere it be blown:
>   But you are more intemperate in your blood
>   Than Venus, or those pamp'red animals
>   That rage in savage sensuality.   (IV. 1)

He no longer notes what she appears (which now seems to have no significance for him) but what she "is":

> O Hero, what a Hero hadst thou been,
> If half thy outward graces had been placed
> About thy thoughts and counsels of thy heart!
> But fare thee well, most foul, most fair! farewell.
> Thou pure impiety, thou impious purity.

If Claudio, so certain of the truth of appearances earlier and, indeed, resting his case on a shadowy set of appearances even here when he denounces Hero at the marriage altar, is sure that the outward show does not sign anything that is inward, there are others present in the scene who take the opposite view. Benedick, for one, noting all the bitterness and cruelty, the seeming deception of Hero, the vindictiveness of her beloved, remarks with cutting irony (at the same time probing the fact that these have been lovers of style): "This looks not like a nuptial."

Another and more serious support of the view that sees more than mere superficiality in the surface of things is also voiced in these painful proceedings:

> *Friar.*   Hear me a little:
> For I have only been silent so long,
> And given way to this course of fortune,
> By noting of the lady: I have mark'd
> A thousand blushing apparitions
> To start into her face, a thousand innocent shames
> In angel whiteness beat away those blushes,
> And in her eye there hath appear'd a fire
> To burn the errors that these princes hold
> Against her maiden truth.

For the Friar, the truth is written on Hero's face. Earlier, in another context, Benedick has declared: "knavery cannot sure hide itself in such reverence" (II. 3), and despite the irony here because of the situation, there is a distinct echo of Ursula's neo-Platonic formula: "Can virtue hide itself? Go to, mum, you are he: graces will appear and there's an end." The play, in other words, is as relentlessly "for" the view that appearances are the whole of reality (or at least accurately reflect reality) as it is "for" the contrary position, that appearances are deceptions, at best, not to be trusted. The question is a complex one, its issues intricate and involved.

One measure of this complexity is the way in which the play's ironies are pushed to greater and greater extremes as the play proceeds. The mechanism of the climactic action is a set of false appearances taken for real ones. Evil, in the person of Don John, makes no direct physical assault upon its victims; it merely manipulates the surface of things, making falsehoods appear true (which, of course, is the primal deed of Satan in the Garden). Real harm, then, is not done by the deceiver, but by the deceived. Similarly, Claudio's good intentions are transformed into an evil deed, not by an active will, but by a transformed sight. It is an altered appearance of reality that transforms his love to hate, his innocence to guilt.

Ironically, however, at the crucial point he fails to save himself and his Hero from their fate, by failing to repeat his procedure and take appearances (now Hero's blushes of innocence as he denounces her) for reflected realities. Thus, Claudio continually misreads the significance of surfaces (which he notes, however, as accurately as circumstances seem to permit) and hence misconceives the realities hidden underneath.

By contrast, the guardians of reality (and hence, innocence) in this play, namely, Dogberry and the Watch, note nothing accurately on the surface of things, but everything underneath. Dogberry is afflicted with a malaprop sense of language that effectively garbles the appearance (and hence the reality) not only of what he says, but of what he hears. It is, in fact, a provident co-ordination of these disabilities that enables Dogberry and the Watch to disentangle appearances and present the main figures of the drama with reality revealed.

The Watch does not "see" the evil that is done (indeed, there is nothing evil to "see") but overhears it. Borachio, explaining to Conrade how he and Margaret had appeared in the window of Hero's chamber and how Don John had deliberately misinterpreted the scene to his brother (Don Pedro) and Claudio, is overheard by the Watch saying:

> . . . chiefly by my villainy, which did confirm any slander that Don John had made, away went Claudio enrag'd . . .   (III. 3)

The Watch mishears what Borachio has said (he thinks that Borachio has said that Don John is a villain) and misinterprets its significance; thus, Borachio is arrested not for the deed which he has confessed, but for slandering Don John, the prince's brother. The Watch has apprehended the right man, for the wrong reason.

This error is corrected, however, when the Watch reports to Dogberry what he has heard. For Dogberry outdoes his subordinate (as is only proper) in seeming incompetence by repeating his error of misnoting what he hears:

> *1 Watch.* This man said, sir, that Don John, the prince's brother, was a villain.
> *Dogberry.* Write down Prince John a villain. . . .   (IV. 2)

Thus, by a process of double negation, the truth is brought to the surface. In other words, a consistently wrong sense of appearances may produce as true a picture of reality as a consistently right one. Indeed, in a world where appearances are deliberately manipulated to imply false realities, only a false sense of (false) appearances can register the truth of events.

In an exquisite and witty moment later on, the tangles of the plot, the confused surfaces and depths, confront each other in an essential way:

> *Dogberry.* Marry, sir, they have committed false report, moreover they have spoken untruths, secondarily they are slanders, sixth and lastly they have belied a Lady, thirdly they have verified unjust things, and to conclude, they are lying knaves.
>
> *Don Pedro.* First I ask thee what they have done: thirdly I ask thee what's their offence; sixth and lastly why they are committed; and, to conclude, what you lay to their charge.
>
> *Claudio.* Rightly reasoned, and in his own division, and by my troth there's one meaning well suited. (v. 1)

Dogberry is speaking about Conrade and Borachio, but what he says is strictly applicable (with the possible exception of the concluding point) to Don Pedro and Claudio. Thus, when Claudio says "there's one meaning well suited," his statement has the double sense of one meaning "put into many dresses" (Dr Johnson) and more importantly, of one meaning appropriate to himself and Don Pedro, as well as to Conrade and Borachio. Thus a diversity of appearances (Dogberry's words) may have one meaning, or significance, while the surface significance of a single meaning (Claudio's statement) may be double.

From this brief rehearsal of themes it is evident that the play raises many more questions than it expects (or can be expected) to answer. But these questions are not primarily raised to be answered. Nor are they simply rhetorical. They serve, rather, an attempt to evoke the special multiple quality of human experience, when probed by an ontological point of view. The issues raised by this attempt cannot be simply summed; but they are implicated (and stunningly so) in a question asked by Dogberry, as he opens the interrogation scene which unravels the entangled matter of the key sequence of misappearances in the play. The question points not only to the moment at hand, but to the larger movement of the play itself:

> Is our whole dissembly appear'd? (IV. 2)

From out of what may be called the play's dissemblage, there emerges more than simple confusion, however. For the play's actions demonstrate not only that the relation between appearance and reality is a sometimes tenuous and always complex one, but also that, because of this uncertainty at the centre of experience, reality is in a large sense what men make it.

II

The whole sequence of misnoted events, and the resulting complex of misdirected responses, in *Much Ado* thus points to the conclusion that phenomenal reality, the reality appearing to men, is the reality that they apprehend—reality, for them, differs with their differing stances towards their condition. For "reality" is no thing, but a *world* —or better still, *worlds*—not single and opaque, but multiple and opalescent.

In *Much Ado,* this protean world character expresses itself as a multiplicity of appearances and responses to appearances; in a sister play, *Twelfth Night, Or, What You Will,* as multiple pleasures and wills to pleasure. Just as a large view enriches experience, so a narrow view is a limiting one, confining life possibilities:

> Dost thou think because thou art virtuous there will be no more cakes and ale?   (Sir Toby)

Therefore, the dialectic of views represented by the two pairs of lovers in *Much Ado* is not resolved by rejecting one for the other, by discarding the romantic perspective for its antithesis. Despite both the failure of Claudio and Hero to navigate a true course, and the corresponding success of the critical pair Benedick and Beatrice, the survival of romance as a viable world-image in the play is assured. It is assured not only by the shallowness of the romantic couple which limits the implications of their actions, but by the final romantic "conversion" of Benedick and Beatrice.

When the deluded Claudio denies Hero at the altar, he is in a sense denying her for the second time. Earlier, when he is falsely led to believe that Don Pedro has usurped his place and wooed Hero for himself, Claudio shows a readiness to be rid of his Hero that calls into question the very existence of a bond of faith between them:

> *Benedick.* . . . the Prince hath got your Hero.
> *Claudio.* I wish him joy of her.
> *Benedick.* Why, that's spoken like an honest drovier. So they sell bullocks. . . .   (II. 1)

To be sure, Claudio speaks out of bitterness. But Benedick's daylight perception is finely tuned to the contours of reality and this is not the first time that he has likened Claudio's attachment to Hero to a commodity relation. (It is worth remembering, in this regard, that the first inquiry Claudio makes about Hero is whether she is Leonato's heir.)

Now, the central fact about a commodity relation is that it is one-sided, the relation of a person to a thing, the one an active possessor, the other a passive possession. The value of the possession, moreover, is inconstant, as is its relation to the person who buys it; a change in its market value may induce him to sell or perhaps even discard it.

Love is the antithesis of his kind of relation. Love is two-sided and therefore, in love, value is not external to, but springs from, the relation. It is not that the lovers possess no merit, no value in themselves, but this value is an active force in each of them that seizes and inspires and draws them together towards possibilities they did not have before. Their relation, therefore, does not remain external to their beings, but they give and hazard all they have. For love involves their very destinies in its "contract," which does not fluctuate with fortune, but is "a world-without-end bargain."

Claudio's denunciation of Hero at the altar reflects the "commercial" one-sided nature of his bond. She has become worthless in his eyes and he has resolved to kill her as a thing of value in all men's eyes, even as she has been killed in his own. Her "death" affects nothing but herself. Claudio loses a jewel, whose value he has been deceived about, but that is all. His world remains intact. In other words, love for him has been no compass, no direction; he has remained outside the sphere of genuine romantic existence, and merely appropriated its rhetoric. Religious symbols provide frills for his experience, but no defining metaphor for his love: to come to the altar for Claudio is simply a ceremonious coming to market.

Claudio's attitude contrasts sharply with that of the true romantic, whose love is a religion and the shattering of whose faith is indeed a shattering of faith, a failure not of mere belief but of the human condition itself, the untuning of cosmic harmonies and the inverting of universal orders. When a romantic like Othello imagines himself betrayed, he speaks in accents of despair which are unmistakable:

> *Othello.* But I do love thee; and when I love thee not
> Chaos is come again—   (III. 3)

"Chaos" is both chaos of his being and the disintegration of his world, a return to primal emptiness. For his love is both the pretext and source of his power, the orientation of his life deed, performed in service of his love.

Othello's love is both origin and destination; he is defined by its compass; its loss deprives him of his role, his "occupation," his very being. In a more than metaphorical sense, his love is the wellspring of his life:

> *Othello.* But there where I have garnered up my heart,
> Where I must either live or bear no life,
> The fountain from which my current runs,
> Or else dries up—to be discarded thence!   (IV. 2)

The venture of love is absolute ("My life upon her faith"). In its binding lovers find the measure of all values; it *bestows* value, and grace:

> . . . were I crown'd the most imperial monarch
> Thereof most worthy, were I the fairest youth
> That ever made eye swerve, had force and knowledge
> More than was ever man's, I would not prize them
> Without her love; for her, employ them all;
> Commend them and condemn them to her service,
> Or to their own perdition.   (Florizel, *The Winter's Tale*, IV. 4)

To find corruption here, is to find everything corrupt; for everything has been measured against this divinity and found wanting. If such prove false, then none can be true. The "fall" inverts all values and significances: divine Desdemona becomes "fair *devil*," the Heaven of Othello's marriage chamber becomes darkest Hell, the deceiver Iago[1] supplants the honest Desdemona as Othello's betrothed ("I am your own forever") and the issue of his marriage is not life, but death.

All this is experience which Claudio remains outside of, throughout his own ordeal. His essential order of reality is not touched by the "exposure" of his beloved. Though he vows to suspect all beauty henceforth, he experiences no elemental crisis, no compulsion to transfer or preserve his faith. The source of his calm lies in the fact that his original bond with Hero was a surface bond, not a faith; his order of reality *originally* remained unaffected (he merely transferred an affection from war to love) and for that reason, through the crisis, can be maintained intact. What destroys Othello, is that he loves Desdemona even after she has "betrayed" him ("Be thus when thou art dead, and I will kill thee and love thee after"). What differentiates Claudio ab-

---

[1] Iago—James—Jacob—the supplanter (Chiappe) [a point made by the late Professor Andrew Chiappe of Columbia University—Ed.].

solutely from Othello is that he has no *need*, even, of preservation; for though he may have fancied Hero, he has never really loved her.

Indeed, Claudio's whole apprehension of Hero has, from its inception, stopped at the surface:

> *Claudio.* In mine *eye* she is the sweetest lady that ever I looked on. (I. 1)

When he first feels that she has betrayed him (with Don Pedro) the lesson he gleans is "Let every *eye* negotiate for itself," and the central deception practised upon him is thus aptly instrumented:

> *Borachio.* . . .I have deceived even your very eyes. . . .   (V. 1)

In the fact that Claudio's sense of Hero is never more than superficial, his suit is revealed to be not love's but fancy's:

> Tell me where is Fancy bred,
> In the heart or in the head?
> How begot, how nourished? . . .
> It is engender'd in the eyes,
> With gazing fed, and Fancy dies
> In the cradle where it lies:
>   Let us all ring Fancy's knell.
>   I'll begin it. Ding, dong, bell.
>   (Song in *The Merchant of Venice,* III. 2)

Benedick and Beatrice, who have been caustically critical of love in general and each other in particular, are brought together by the second major elaborate deception of the play. Separately, they are made to overhear reports concerning the transformation of each other's feelings. Benedick's first reaction to the intelligence that Beatrice loves him is: "This can be no trick." Since he thinks that he is overhearing a conversation which was not meant for his ears (he is concealed), the very suspicion that it might be a trick is stronger than the denial itself and thus, as an intuition, is particularly well tuned. The irony of the situation is, of course, double. For the deception of Benedick is only a surface deception. Benedick and Beatrice really are in love, and Benedick's readiness to accept the intelligence of the "deception" reveals the extent to which his perspective comprehends the possibility of this love. He is not surprised to find that his severest critic is his truest admirer. For the one is a function of the other, and Benedick and Beatrice have all along, and in the deepest sense, been concerned with each other. Benedick, in the end, is right: this can be no trick.

Benedick's intuition into the "truth" of his experience rests not only on deep self-knowledge, but on a corresponding awareness of the character whom he must rely on to perceive and interpret events for him. It is, in fact, Hero's father Leonato who confirms the (false) reports of the evidence of Beatrice's love for Benedick. Noting from whose lips the intelligence comes, Benedick concludes: "Sure, knavery cannot hide itself in such reverence." And he is right. This is not knavery, but a well-meaning trick based on an accurate perception about himself and Beatrice, and motivated by a benevolent intention towards them.

By contrast, Claudio depends for his "intelligence" on the bastard Don John, who is unscrupulous and has reason to wish to do harm to Claudio and the Prince. Indeed, when the Friar suggests that the three who claim to have "seen" Hero's perfidy have been deceived in some way, Benedick quickly locates the source of the deception. He notes that two of them (Claudio and Don Pedro):

> . . . have the very bent of honour
> And if their wisdoms be misled in [regard to Hero]
> The practice of it lies in John the bastard,
> Whose spirits toil in the frame of villainies.   (IV. 1)

Thus, the very attention to concrete details which characterizes the satiric vision, enables Benedick to sort out appearance from reality; and, conversely, it is Claudio's romantic penchant for projecting his vision away from the concrete, for adorning the mundane with fantastic suits, that prevents him from seeing truly.

Notwithstanding that they have begun their relation within the context of a complex set of appearances (they have been mutual antagonists—they have merely heard reports of each other's affection), Benedick and Beatrice come quickly to the firm ground of real engagement, and to the mutual confrontation which attends that arrival.

Their declarations of love follow on the disgrace of Hero and are precipitated by Beatrice's distress over this event, which Benedick seeks to salve. Benedick's first avowal is couched in accents which, while certainly romantic, remain true to his original mode:

Benedick. I do love nothing in the world so well as you. Is that not strange?
                                                                            (IV. 1)

This formulation in the negative provides a kind of cover for his nakedness, giving an impression of reasonableness that the more orthodox—"I love you more than anything"—would not have. But Beatrice remains non-committal.

Benedick. By my sword, Beatrice, thou lovest me.
Beatrice. Do not swear and eat it.

He cannot compel her into love. It is her grace to give him, and he knows it. He can swear for her, but confirm it only for himself:

> *Benedick.* I will swear by it that you love me, and I will make him eat it
> that says I love not you.
> *Beatrice.* Will you eat your word?
> *Benedick.* With no sauce that can be devised to it. I protest I love thee.

This naked avowal satisfies Beatrice and she responds, wittily at first; then openly:

> *Beatrice.* I love you with so much of my heart that none is left to protest.

Her gift is grace. His answer acknowledges it and seals, moreover, the basic loss of his cherished singleness. He has given his heart away:

> *Benedick.* Come, bid me do anything for thee.

This is the true romantic gesture, the confirmation in deed of the transcendent value of the relation. Love here manifests itself as a real force in the world, as the absolute source of direction, as ultimate significance. She takes him at his word:

> *Beatrice.* Kill Claudio.

Love is immediately the basis for decision over life itself. Beatrice's love for Hero determines her decision; Benedick's love for Beatrice must determine his. Her command, accordingly, is absolute, as absolute as the bond which they have sealed, as absolute as its horizon, which is death.

> *Benedick.* Ha! Not for the wide world.
> *Beatrice.* You kill me to deny it. Farewell.

The bond that is between them is a word only, and can never be more than a word. Benedick has seemed to give content to this word—"Come, bid me do anything for thee"—but now he withdraws what he has given, *breaks* his word and, in Beatrice's eyes, its substance. For love has "substance" only in so far as it continually manifests itself in action, in the actual being of the lovers. Thus, even as Benedick empties his word of its content, he denies himself the power to manifest what he feels and intends, and so to make Beatrice experience his love:

> *Benedick.* Tarry, sweet Beatrice.
> *Beatrice.* I am gone though I am here. There is no love in you. Nay, I pray
> you let me go.
> *Benedick.* We'll be friends first.

But a bond is manifested by what it can stand against and by the actions it will beget:

> *Beatrice.* You dare easier be friends with me than fight with mine enemy.
> *Benedick.* Is Claudio thine enemy?
> *Beatrice.* Is he not approved in the height a villain that hath slandered, scorned, dishonoured my kinswoman? Oh that I were a man! What, bear her in hand until they come to take hands,[2] and then, with public accusation, uncovered slander, uninstigated rancour—Oh, God, that I were a man! I would eat his heart in the market place.
> *Benedick.* Hear me, Beatrice—

But she will not listen and he cannot persuade her. Moreover, he has, himself, become linked with Claudio's treachery; for Claudio's nobility, grace, and honour have revealed themselves to be mere surfaces and forms, even as Benedick's vow seems to be mere words:

> *Beatrice.* Princes and Counties! Surely a princely testimony, a goodly Count, . . . Oh, . . . that I had any friend would be a man. . . . *But manhood is melted into courtesies, valour into compliment, and men are only turned to tongue, and trim ones too. He is now as valiant as Hercules that only tells a lie, and swears it. . . .*

Value has become a mere dressing, an empty name, easily come by, since no one is ready to confirm its reality in action. If Benedick really loves Beatrice, then he must prove the substance of his vow:

> *Benedick.* Tarry, good Beatrice. By this hand, I love thee.
> *Beatrice.* Use it for my love some other way than swearing by it.[3]

Now Benedick must confront the meaning of his engagement. In a world of shifting appearances, where what is true is elusive, and to be converted is to see with different eyes, he has entered upon a trust that is absolute, that cannot survive the denial of its claim. No end can be served by disputing the meaning of appearances with Beatrice; she loves Hero and believes in her, and Hero's presumed death is a claim that she must answer. By his engagement to Beatrice, Benedick has become involved in her order of reality. Now he must depend on her word; moreover, he must confirm his own word to her with a deed. Love is his compass.

[2] I.e. in the hand-fast of marriage.
[3] Another play on hand-fast.

> *Benedick.* Think you in your soul the Count Claudio hath wronged Hero?
> *Beatrice.* Yea, as sure as I have a thought or soul.

In the end, it will be as it must be. The toughness of Beatrice's faith will be more impressive testimony for him than the evidence of Claudio's eyes. And he will be right.

> *Benedick.* Enough, I am engaged, I will challenge him. . . . *As you hear of me, so think of me.*

Here the union, which is a double union, is sealed. Love binds their formerly separate lives in a single destiny, and, in the deed of love itself, appearance and reality become one. Here they achieve the plane of romantic action: they are religiously committed within a completely human frame.

It is, moreover, their critically imaginative awareness that has made this achievement possible. Being satirically disengaged, they have been able to dwell imaginatively in many realms, and this imaginative grasp of experience and its possibilities has prepared them beforehand for the seriousness of the romantic venture. Benedick, for example, has been a master of the romantic mode long before he has been conscious of his love:

> *Benedick.* O, she misus'd me past the endurance of a block! . . . If her breath were as terrible as her terminations, there were no living near her; she would infect to the north star. . . . *Will your Grace command me any service to the world's end? I will go on the slightest errand now to the Antipodes that you can devise to send me on; I will fetch you a toothpicker now from the furthest inch of Asia, bring you a length of Prester John's foot, fetch you a hair of the great Cham's beard, do you any embassage to the Pigmies, rather than hold three words, conference with this harpy.* (II. 1)

Here is the romantic quest with a vengeance, albeit as a quest to *avoid* the lady. Moreover, the hyperbole employed by the outwitted Benedick goes far beyond any purpose of merely mocking the romantic ideal of service. There is a frustration in its extremity whose counter-side is devotion. This serves to recall that Benedick has never failed to appreciate Beatrice's graces and that he has early declared to Claudio that were Beatrice not "possessed with a fury," she would exceed Hero "as much in beauty as the first of May doth the last of December."

Indeed, the "conversion" of Benedick and Beatrice, in one sense, has been simply a reversal of direction, not an adoption of a new order. The religious mode of committed existence has been their mode all the while. Benedick is renowned for both his great valour and his unwavering honour. If Benedick and Beatrice have lacked faith, it has been faith that springs from a single mortal human being, not faith as such. They have not been ignorant, moreover, as Claudio has, of the nature of love's bond. On the contrary, the very basis of their resistance to the notion of human love was their precise knowledge of what was at stake. They knew this in the only way they could: *imaginatively*. What they did not know was that such a religion could be seriously viable when sustained by two people with eyes for each other's frailties. In the end, they have to acknowledge that such a faith can sustain itself. But they do so without the self-delusions, without the postulation of a divine image, upon which Claudio's "love" depends:

> *Benedick.* . . . And I pray thee now, tell me for which of my bad parts didst thou first fall in love with me?
> *Beatrice.* For them all together, which maintained so politic a state of evil that they will not admit any good part to intermingle with them. . . .
>   (v. 2)

Beatrice here aims a superbly self-conscious barb at the illusions of lovers. Benedick not only has bad attributes for which she loves him, but his badness is perfection; it admits no single good element to adulterate its purity. In response to this jibe, Benedick drily observes:

> *Benedick.* Thou and I are too wise to woo peaceably.

Their critical realism gives to the bond that is between them a resilient strength. Wittily, they can face the absurdity of love and its wisdom: that it is between two people, frail and fallible, whom mortality never fails to touch; that precisely in this—that men are born to die, and love is for them creation and renewal—lies its necessity and its grace.

Beatrice and Benedick discover what only the sceptic empowered by imagination and ever open to the possibility of commitment can discover: that love is indeed an idea out of imaginative fiction, but that like all ideals it need not remain merely a potential. The real commitment of two lovers may yield love a substantiality and permanence that no dream can have. In this perception Beatrice and Benedick make their way to a central, commonplace, paradoxical truth: that love, in its reality, is romance.

Benedick and Beatrice have multiple perspectives; they are aware of worlds of being and have the power and the will to embrace them, to render substance to what otherwise would remain mere names. In a marvellous metaphor, this restoration of content to form is a restoration for themselves as well. The sacrament of marriage gives them to themselves: they are Benedick—*benedictus* and Beatrice—*beatus,* the blessed. And, in celebration of this happiness and harmony, Benedick undergoes a final conversion to music:

> *Benedick.* . . . Let's have a dance ere we are married, that we may lighten our own hearts, and our wives' heels.
> *Leonato.* We'll have dancing afterward.
> *Benedick.* First, of my word; therefore play, music. (v. 4)

Ritual and Insight

by Francis Fergusson

When Shakespeare wrote *Much Ado About Nothing* he had lost none of his skill as a maker of plots; on the contrary, he had attained further mastery in the ten years or more since the writing of *The Comedy of Errors*. There are three main narrative-lines: that of Claudio, Hero, and the wicked Don John; the connected story of Dogberry and the Watch; and the contrasting story of Beatrice and Benedick, all interwoven with clarity and apparent ease. But in this play Shakespeare uses the plot for a further and deeper end. Each of the three narrative-lines has its own humor, and by the interplay of the three a more general vision of man as laughable is suggested: a vision which is at once comic and poetic.

The story of young Claudio and Hero caught in Don John's wicked schemes was Shakespeare's starting point, and the somewhat casual framework of the plot of the whole play. He had read this story in Bandello's version, *Timbreo di Cardona,* the story of a girl unjustly accused of adultery. This tale, though it ends happily, is not very funny in itself, and Shakespeare does not so much avoid its painful and pathetic aspects as absorb them in his more detached comic vision. The scene in the church, when poor Hero is wrongly accused and her father Leonato loudly laments, may be played for a "tragic" effect, but that I think would not be quite right. The audience knows that it is all a mistake, and it is by that time accustomed to smile at Claudio, an absurdly solemn victim of young love's egoism. When he first appears he tries to tell the Duke what the Duke knew already: his all-important love for Hero. He glumly decides that the Duke, wooing Hero in his behalf, has stolen her, and so is wrong again. Beatrice labels him for us: "glum as an orange, and something of that jealous complexion." His false accusation is his third mistake: we must sympathize, but at

"Ritual and Insight" [*Editor's title*] *by Francis Fergusson. From* The Human Image *in Dramatic Literature (New York: Anchor Books, Doubleday & Company, Inc., 1957), pp. 150–57. Copyright © 1957 by Francis Fergusson. Reprinted and retitled by permission of the author.*

the same time smile, at this final instance of his foolishness. The whole Claudio-Hero story is comic in itself and in its own way, but to understand what Shakespeare meant by it it is necessary to think of it in relation to the two other stories which unfold in alternation with it.

Dogberry and the Watch are closely connected with the Claudio story, which requires someone to uncover Don John's plot, but Shakespeare developed this element into a farcical sequence with its own tone and interest. At the same time he uses it to lighten the catastrophe at Hero's wedding, and the character of Don John: we cannot take a villain seriously who can be apprehended by Dogberry. Dogberry is not suffering the delusions of young love, like Claudio, but those of vanity and uncontrollable verbosity. His efforts to find his way, with lanterns, through the darkness of the night and the more impenetrable darkness of his wits, forms an ironic parallel to the groping of the young lovers through their mists of feeling. Dogberry also has his version of the underlying mood of the play—that of a leisurely and joyful ease, such as we attribute to Eden or the Golden Age. In Dogberry this infatuated leisureliness, this delusion that nothing terrible can really happen, takes the form of interminable verbalizing while the evil plot hatches and the villains lurk uncaught.

The story of Beatrice and Benedick's self-tormented love affair is entirely Shakespeare's creation. He seems to have felt the need of that pair's intelligence and agility to ventilate Claudio and Hero. We should tire quickly of Claudio's total submersion in love if Benedick were not there, pretending to be too intelligent for that. Hero, who can only sigh and blush, would be too soggy without Beatrice, who can only make sharp remarks, pull pigtails, and stick her tongue out at the boys. But the two contrasting stories together suggest a vision of early infatuation—provided we don't take Shakespeare's characters more seriously than he intended—which is both deeper and more comic than the victims themselves can know.

Beatrice and Benedick are notoriously hard to act on the modern stage, especially in the first two acts, where they indulge in so many quibbles and conceits in the taste of their times. There is no use trying to make the verbal jokes funny; but I am not sure that Shakespeare himself took them seriously as jokes. I once had the pleasure of seeing John Gielgud and Pamela Brown act several of the Beatrice-Benedick scenes. They "threw away" the words, or even, at moments, made fun of their far-fetched elaboration, and by this means focused their audience's attention on the noble, silly, intelligent and bewildered *relation* of the two—a relation as agile, musical, and deeply comic as that of Congreve's reluctant lovers, Mirabel and Millamant. I feel sure that this approach to the play is right: its surfaces, its literal words, charac-

ters and events, are not to be taken seriously: the point is in the music
of unseen motivation, in the fact that it *is* unseen by the characters
themselves—and that all the fun and folly plays against a background
of mystery.

The main Claudio-Hero-Don John intrigue is also not to be taken
too seriously, as though it were the point of the play: Shakespeare gets
it under way casually, after the underlying mood of the play as a
whole, and its "action" of elaborate play, or leisurely enjoyment, has
been firmly established. The opening scene, in which Leonato's household prepares to celebrate the return of the Duke, Benedick and
Claudio from their comic-opera war, tells us what the play is really
about: it is a festive occasion, a celebration of a certain evanescent but
recurrent human experience. The experience is real in its way, all may
recognize it, but under its spell everything the characters do is much
ado about nothing. The progress of the underlying action of the play
as a whole is therefore marked by a series of somewhat dreamy and
deluded festive occasions. The first of these is Leonato's masked ball, in
Act II, a visible and musical image of the action. Then comes Dogberry's nocturnal and incomprehensible charge to the Watch: a farcical
version of the theme. The fourth act consists chiefly of the marriage
which turns out to be no marriage at all, but a bad dream. In the fifth
act there is Claudio's funeral tribute to Hero, by night, at her supposed tomb; but this is a funeral which is no funeral, corresponding
to the marriage which was no marriage. After that pathetic and comic
expiatory rite, daylight returns, the torches are put out, and we are
ready for the real and double marriage, in daylight, with the ladies
unmasked at last, which ends the play in dance and song.

We are just beginning to understand the technical value of the
"ceremonious occasion" as an element of plot, though it has been used
in countless ways from Aristophanes to Henry James. When people
assemble for a ceremonious occasion (whether it be the festival of
Dionysos or one of James's thorny tea parties) they must abate, or
conceal, their purely individual purposes, and recognize the common
concern which brings them together. A dramatist may use the festive
occasion, therefore, to shift his audience's attention from the detail of
the literal intrigue to some general plight which all more or less unwit-
tingly share. All are social and political animals; all must suffer spring,
mating, and death. Ceremonious occasions are especially useful to
dramatists who are seeking poetry, which, as Aristotle remarked, is
concerned with something more general than the particular facts, the
unique events, of human life. The point—the comic point—of *Much
Ado*—is poetic in that sense, and hence it is the festive ensemble scenes

which most clearly adumbrate the basic vision of the play. In this respect the plot of *Much Ado* contrasts sharply with that of *The Comedy of Errors*. The point of that play lies precisely in the unique situation of mistaken identity, and in the strings of absurd events which quickly follow from it. An "occasion" of any kind would break the tight concatenation of *contretemps;* and that Shakespeare is careful to avoid doing until he is ready to end the whole play.

One might say that *Much Ado* presents a comic vision of mankind which is also poetic, while the purpose of *The Comedy of Errors* is closer to that of the professional vaudevillian, who gauges his success by clocking the laughs: the provoking of thoughtless mirth, an almost reflex response. The difference between the two plays is clearest, perhaps, when one reflects that both are concerned with mistaken identity, but in *The Comedy of Errors* the mistake is simply a mistake in fact, while in *Much Ado* it is a failure of insight, or rather many failures of different kinds by the different characters.

Shakespeare accomplishes the *dénouement* of *The Comedy of Errors* in one swift scene. It is not difficult to correct an error in fact: it may be done instantly by providing the right fact: and as soon as both pairs of twins are on stage together, the error is gone. But correcting a failure of insight is a most delicate and mysterious process, which Shakespeare suggests, in *Much Ado*, in countless ways: through the symbolism of masks, night, and verbal ambiguities, and in peripeteias of his three variously comic subplots.

The farcical efforts of Dogberry and Verges never deviate into enlightenment. They learn as little as the characters in *The Comedy of Errors:* but, like them, they do stumble eventually upon the right fact: they manage to apprehend the villains and convey that fact to Leonato.

Claudio, with his dark fumes of love, has a long way to go before he can see anything real. After his false wedding Shakespeare puts him through a false and painful challenge from his best friend, Benedick, and then the mocking (but touching) mummery of his visit to Hero's empty tomb. Even then the audience learns more from Claudio's masquerade-like progress through the maze than he does himself.

Beatrice and Benedick come the closest, of all the characters, to grasping the whole scope of the comic vision which the play slowly unfolds. But even after their friends have tried to kid them out of their frightened vanity during the first three acts, it takes most of the fourth and fifth acts, where all the painful things occur, to bring them to conscious acceptance of their absurd selves, each other, and their love. It is the fiasco of Claudio's first attempt at marriage which marks the crucial turn in their relationship:

> *Benedick.* Lady Beatrice, have you wept all this while?
> *Beatrice.* Yea, and I will weep a while longer.

and a little later:

> *Benedick.* I do love nothing in the world so well as you. Is not that strange?
> *Beatrice.* As strange as the thing I know not. It were as possible for me to
> say I love nothing so well as you; but believe me not; and yet I lie not;
> I confess nothing. . . .

In this exchange the love-warmed final scene of the play is foreshadowed, but the misfortunes of Claudio and Hero, which here bring Beatrice and Benedick near together, immediately carry them apart again. Benedick has to challenge Claudio, and that boy's delusions have to be repented and dispelled, before Beatrice and Benedick can trust their intuition of love, or accept it fully and in good conscience. I do not attempt to follow the subtle shifts in their relationship which Shakespeare suggests, in a few quick, sure strokes, during the fifth act. But it is Beatrice and Benedick who dominate the final scene:

> *Benedick.* Soft and fair, Friar. Which is Beatrice?
> *Beatrice (unmasking).* I answer to that name. What is your will?
> *Benedick.* Do not you love me?
> *Beatrice.* Why no, no more than reason.
> *Benedick.* Why then, your uncle and the Prince and Claudio have been de-
> ceived; they swore you did.
> *Beatrice.* Do not you love me?
> *Benedick.* Troth no, no more than reason.
> *Beatrice.* Why then my cousin, Margaret, and Ursula are much deceived,
> for they did swear you did.

(Claudio and Hero produce love letters from Benedick and Beatrice to each other)

> *Benedick.* A miracle! here's our own hands against our hearts. Come, I will
> have thee; but, by this light, I take thee for pity.
> *Beatrice.* I would not deny you, but by this good day I yield upon great
> persuasion, and partly to save your life, for I was told you were in a
> consumption.
> *Benedick.* Peace; I will stop your mouth.

In this scene the main contrasting themes of the play are brought together, and very lightly and quickly resolved: marriage true and false, masking and unmasking, the delusion and truth of youthful love. The harmonies may all be heard in Beatrice's and Benedick's words. The exchange is in prose, but (like the prose of Leonato's masked ball) it has a rhythm and a varied symmetry suggesting the formality of a dance figure. The key words—love, reason, day, light, pity, peace— make music both for the ear and for the understanding as they echo back and forth, deepening in meaning with each new context. The effect of the scene as a whole is epitomized in Beatrice's and Benedick's heavenly double-take: their foolish idiosyncrasy is clear, but some joyful flood of acceptance and understanding frees them, for the moment, and lifts them beyond it. Is this effect "comic"? I do not know; I think it is intended to bring a smile, not for the windup of this little plot, but for the precarious human condition.

When one reads *Much Ado* in the security of one's own room, indulging in daydreams of an ideal performance, it is possible to forget the practical and critical problems which surround the question of the play's viability in our time. But it must be admitted that high school productions are likely to be terribly embarrassing, and I do not even like to think of the play's pathetic vulnerability on Times Square. The play demands much from its performers, almost as much as Chekhov does. It demands a great deal from its audience: a leisurely and contemplative detachment which seems too costly in our hustled age. Perhaps Shakespeare should be blamed for all this: if *Much Ado* does not easily convince us on the contemporary stage, perhaps we should conclude, as Eliot once concluded of *Hamlet,* that it is an artistic failure. But on that principle we should have to rule out a great deal of Shakespeare. It was his habit, not only in *Hamlet* and *Much Ado,* but in many other plays, to indicate, rather than explicitly to present, his central theme; and to leave it to his performers and his audience to find it behind the varied episodes, characters, and modes of language which are literally presented. Everything which Shakespeare meant by *The Comedy of Errors* is immediately perceptible; the comic vision of *Much Ado* will only appear, like the faces which Dante saw in the milky substance of the moon, slowly, and as we learn to trust the fact that it is really there.

Forgiving Claudio

by Robert Grams Hunter

The comedies of forgiveness, like all romantic comedies, celebrate the love of man for woman, but by their very nature they are also celebrations of that variety of love which the Bible calls charity. Romantic love is celebrated as a source of happiness for the man and woman who love each other, and as the socially acceptable form of the force upon which the continued existence of society depends—sexual desire. Charity is celebrated also as a virtue in itself and as a social necessity, for the comedy of forgiveness proclaims that human nature is such that any social structure must finally depend on mutual forbearance—on forgiving and being forgiven. More than that, romantic love itself turns out in these plays to be dependent upon the virtue of charity. The love of man for woman (but not of woman for man) is seen as too frail an emotion to sustain the pressures that are frequently put upon it. Man's love fails, and woman must charitably forgive the failure.

The error of the *humanum genus* figure in four of our comedies consists in this failure of love. As a result of deception or lack of self-knowledge (or both), the romantic heroes fail the women who love them, and by this failure hatred gains the ascendancy within their minds. As a result, the fabric of the play's society is threatened by strife, and love and order are finally restored only after a revelation of truth and a consequent repentance and forgiveness. This is the pattern of *Much Ado, All's Well* (though Bertram is something of a special problem), *Cymbeline,* and *The Winter's Tale.*

* * *

But we must now ask ourselves precisely what Claudio, as *humanum*

"Forgiving Claudio" [Editor's title] by Robert Grams Hunter. From "Much Ado About Nothing," in Shakespeare and the Comedy of Forgiveness (New York: Columbia University Press, 1965), pp. 93, 98–105. Copyright © 1965 by Columbia University Press. Reprinted and retitled by permission of the publisher.

genus, is guilty of. Clearly Shakespeare did not intend that we should blame him for believing in the tableau presented for his benefit by Don John. Don Pedro, whom we are meant to like and admire, is equally taken in and the very similar charades got up for the benefit of Beatrice and Benedick demonstrate that even the wittiest and wariest can be fooled by appearance. To fall victim to illusion in this play is to rank oneself with the good, for in *Much Ado* as in *Othello,* only the villain is never fooled. But if, in Claudio's own words, he "sinned . . . not, / But in mistaking," (v. i. 283–4) then clearly he is more to be pitied than censured and Don John was the cause of it all.

The critical reaction to Claudio has fallen generally into one of two categories. Traditionally, he has been seen, in Andrew Lang's excellent phrase, as "the real villain as well as the *jeune premier* of the piece," [1] and the critics have assumed that Shakespeare simply botched the job of creating a standard *jeune premier.* In recent years, however, a number of writers on the play have taken a quite different approach to the character. Recognizing correctly that Claudio cannot be blamed for being deceived by Don John, they move to the further conclusion that he cannot be blamed for anything. According to T. W. Craik, "Most critics blame Claudio's credulity and/or his public repudiation of Hero. . . . In reality, Claudio is exonerated, chiefly by the facts that Don John (as villain) draws all censure on himself and that Don Pedro (hitherto the norm, the reasonable man) is also deceived." [2] Let us look at that repudiation and see to what extent Claudio can be so exonerated.

In the repudiation scene we see a girl whom we know to be entirely innocent and loving met at the marriage altar by the man who is supposed to become her husband and forced to listen to—among others—the following sentiments from the man she loves:

> There Leonato, take her backe againe,
> Giue not this rotten orenge to your friend.
>
> . . . . .
>
>   would you not sweare
> All you that see her, that she were a maide,
> By these exterior shewes? But she is none:
> She knowes the heate of a luxurious bed:
> Her blush is guiltiness, not modestie.
>
> . . . . .

---

[1] Lang, *"Much Ado About Nothing," Harper's Magazine,* LXXXIII (September, 1891), 502.

[2] *"Much Ado About Nothing," Scrutiny,* XIX (October, 1953), 314.

But you are more intemperate in your blood,
Than Venus, or those pampred animalls,
That rage in sauage sensualitie.   (IV. i. 32–62)

It may ultimately be Don John's fault that Claudio thinks and says
these things, but Claudio thinks and says them. His passionate outrage
causes him to behave cruelly, and we, especially since we know that
Hero is innocent, are repelled by his cruelty. To be sure there are good
reasons for his behavior—more and better excuses than Shakespeare
gives to any of his other *humanum genus* figures. Claudio is taken in
by a villain, and Don Pedro is also deceived. As a result of Don John's
trick, Claudio is convinced that his honor has been endangered, and
he determines not simply to avoid what he thinks is a trap set for him,
but to revenge himself for almost being duped. To some extent,
Claudio's cruelty to Hero is inspired by social considerations. He and
Pedro think Hero has misbehaved and are resolved that she shall be
punished for her conduct. Some critics would have us believe, therefore, that Claudio's outburst against Hero in the church scene represents nothing more than the emotion proper to a proper Elizabethan
whose proprieties have been outraged, and that the denunciation of
Hero would have been automatically approved by an Elizabethan
audience: "Had [Claudio], as in the sources, quietly told Hero and her
father that the nuptial rites could not be celebrated, the Elizabethan
audience would have felt that justice had not been done." [3] But surely
this statement contradicts itself, for if the young man's behavior is
proper in the sources, why should it not have remained proper in
Shakespeare's play? An interesting point has been raised, however.
Shakespeare could have done the whole thing differently if he had
wanted to.

If Shakespeare had wanted to create a Claudio who was in all respects
an exemplary *jeune premier*, he might have done so easily. Any hack
dramatist could dash off a scene in which a manly but deeply wounded
Claudio asks his friend and prince to explain to Leonato why the
marriage cannot take place. Shakespeare does not do so because he
does not want to do so. Shakespeare wants to show us in Claudio a man
in love whose love has not been able to survive the severe strain to
which it has been subjected. In the terms of Sonnet 116, Claudio has
admitted impediments to the fulfillment of his love. He has bent with
the remover to remove without realizing that the remover is not Hero
but Don John. The real source of Claudio's outburst against Hero is
not outraged propriety. Nor is it, as C. T. Prouty has suggested,[4] the

[3] Page, "The Public Repudiation of Hero," *PMLA*, L (September, 1935), 743.
[4] *The Sources of "Much Ado About Nothing,"* p. 62.

chagrin of a man who believes that the value of the goods offered him in a *mariage de convenance* has been misrepresented. Claudio's brutal denunciation of Hero is the result of his former love for her—a love that has been transformed into a hatred all the more intense because it was formerly love. Though his hatred has been released by a villain, it is an ugly emotion and Claudio expresses it in as ugly a way as possible so as to ensure that we will dislike him for it. Shakespeare wants us, for the moment, to dislike Claudio intensely, and later, when Beatrice ends one of her magnificent tirades against this "goodly Count Comfect" with the finely Sicilian "O God that I were a man! I woulde eate his heart in the market place," (IV. i. 308–9) she is expressing and consequently relieving some of the emotion that we have stored up against her enemy.

Yet it is this speech that begins to rebuild the character of Claudio a bit. The very vehemence of the attack reminds us that it is made on false grounds, that Claudio is not precisely the villain Beatrice takes him for. We have already seen that Beatrice's attack on Claudio resembles Claudio's on Hero. There are, however, differences that are certain to affect decisively the audience's reactions to each outburst. Beatrice's attack is basically a defense, a counterattack in which she tries to save her cousin's honor, and she is inspired more by love for Hero than by hatred for Claudio. Most importantly, the object of the attack is not a person for whom Beatrice has ever felt any unusual affection. Claudio's tirades against Hero are very different. They spring entirely from a love-destroying hatred and they are a direct attack upon the very heart of the comic mystery—romantic love. They are an expression of the hatred that love contains, and the terms which Claudio uses betray a revulsion against sexuality itself. One of the defenders of Claudio's treatment of Hero has approved of this aspect of it on the grounds that it "shows an abhorrence of such carnality," the effect of which is to "idealize Claudio even as he denounces the innocent Hero." [5] But is an abhorrence of carnality really likely to idealize a romantic hero? Even in tragedy the disgust of Hippolytus with the idea of sexuality makes us uncomfortable, and Hamlet's tirade against Ophelia provides his least heroic moment. In romantic comedy, whose purpose is to idealize and celebrate sexual love, such "abhorrence" is distinctly out of place.

In fact, some of Claudio's worst offenses are those against the form and spirit of comedy. Had his hatred and denunciation of Hero taken place in reality (as his defenders tacitly assume they have) most men

[5] Neill, "More Ado About Claudio: An Acquittal for the Slandered Groom," *Shakespeare Quarterly*, III (April, 1952), 97.

would probably find them in some degree offensive. Occurring as they do in a comedy, they become very difficult to pardon. We have—we members of the audience—gathered together, after all, in the expectation of enjoying a spectacle designed to convince us, however momentarily, that beauty exists and that happiness and love are goals possible to attain. The wedding to which *Much Ado* brings us in Act Four is precisely one of those festivities with which comedy should properly end. By preventing us from vicariously enjoying the pleasures symbolized by that ceremony, by rudely refusing us the spectacle of love and happiness attained, Claudio aligns himself with those interrupters of festivity which comedy has always contained and whose defeat has always provided a large part of the comic pleasure. Claudio, in terms of comic character functions, is transformed from a romantic hero into a barrier figure—a personified impediment to the happiness toward which comedy moves.[6]

The writers of comedy discovered early that the pleasures of arriving at a happy ending were all the more intense if the journey had been difficult, and Plautus and Terence (presumably having learned from Menander) became adept at the devising of barriers to felicity. These (usually personified) barriers, though they exercised the agility of the characters on stage, did not demand anything more of the audience than an amused attention. The Shakespearean comedies we are discussing are different. The fact that here the barriers have been placed within the minds and feelings of the characters who are destined for happiness means that we must participate emotionally in the achievement of the happy ending. An on-stage happy ending in these plays is dangerously easy to attain. The forgiveness of the offenders by the offended is all that is needed. But if we in the audience are to participate in that felicity, we must also participate vicariously in the means to it. *We* must pardon the offenders. If we cannot, the play does not, for us, end happily, and we are denied the comic experience.

Having allowed his *humanum genus* figure to offend both the moral and aesthetic sensibilities of the audience, Shakespeare must now set about cajoling us into forgiving Claudio his trespasses. The two methods he uses to do so are excuse and penance. By personifying the origin of evil in Don John, Shakespeare has provided a scapegoat upon whom to heap Claudio's misdeeds at the end of the play. By allowing Don Pedro to share Claudio's errors, Shakespeare lightens them. He has taken unusual care in *Much Ado* to provide his *humanum genus* with a strong defense against the enmity of the audience. In the rest of his comedies of forgiveness, he is by no means so tender toward his

[6] See Frye, *The Anatomy of Criticism*, p. 174.

erring heroes. But despite the excuses with which he is provided, a residue of guilt remains with Claudio, and he must expiate it. Through his hatred and cruelty, he has enabled evil to enter the world of comedy. To deserve our forgiveness, he must be put through a process of penance.

That process resembles, in its elements, the sacrament of penance as it is described by Aquinas. Claudio experiences contrition, confesses his sin, and agrees to make satisfaction. His contrition is the result not, as the friar had hoped, of the news of Hero's death, but of the confession of Don John's accomplice, Borachio, who makes it clear that Claudio's hatred and cruelty have been the product of a terrible illusion. When Borachio's speech has run like iron through his blood, Claudio turns to Leonato for the imposition of punishment:

> Impose me to what penance your inuention
> Can lay vpon my sinne, yet sinnd I not,
> But in mistaking.   **(v. i. 282–84)**

Don Pedro joins him:

> By my soule nor I,
> And yet to satisfie this good old man,
> I would bend vnder anie heauy waight,
> That heele enioyne me to.   **(v. i. 284–87)**

The weight is not extraordinarily heavy. Leonato asks them to make their confession public:

> Possesse the people in Messina here,
> How innocent she died, and if your loue
> Can labour aught in sad inuention,
> Hang her an epitaph vpon her toomb,
> And sing it to her bones. . . .   **(v. i. 290–94)**

Further "satisfaction" will be Claudio's alone. In order to redress the wrong he has done to the honor of Leonato's family, he must marry Leonato's niece. In the comedy of forgiveness, however, contrition and confession are usually enough. The sins of the *humanum genus* figure are revealed to have failed of their full effect. Claudio's outburst of venomous hatred has injured Hero, but has not destroyed her. The death of Hero is the last of the play's illusions and, like the double deception of Beatrice and Benedick, its effects are beneficent. When Hero unveils, Claudio awakens from the nightmare that has been imposed upon him by the wisdom of the friar. With that awakening, the happiness within the play is complete. The force of love is

once again ascendant over hate. The comic world has been restored to
its natural condition. Within the play, Claudio is unreservedly forgiven. When the friar says of Hero, "Did I not tell you shee was
innocent?" Leonato replies:

> So are the Prince and Claudio who accusd her,
> Vpon the errour that you heard debated.   (v. iv. 2–3)

Whether or not the audience joins wholeheartedly in this charity is
definitely less certain. It is clear, I think, that Shakespeare meant us to,
that he wished us to pass on to the concluding dance with minds un-
troubled by doubts as to Claudio's worthiness, and it seems likely that,
in his own time, he achieved his desired effect. M. C. Bradbrook betrays
the usual modern uneasiness about Claudio's contrition when she says,
"Claudio cannot now be . . . allowed more than a pretty lyric by
way of remorse." [7] It is possible that an Elizabethan audience would
have found more, and what they found might have inspired them
to a stronger sympathy with the offender. Shakespeare presented them
with the spectacle of a man falling victim to false appearance and, as
a result, becoming possessed by the force of hatred, in other words,
with the spectacle of a man behaving like a man. In fact, Claudio's
crime is being human. There is no worse offense, but surely an audience of human beings should not indulge itself in a complacent sense
of superiority to such an offender. The Elizabethan dramatic tradition
would have disposed *Much Ado*'s first audience to see in Claudio, and
figures like him, images of its own frailty, and the religious connota-
tions which Shakespeare gives to the penance of Claudio would have
emphasized that resemblance.

[7] Bradbrook, p. 187.

The Role of Wit in *Much Ado About Nothing*

by William G. McCollom

Much Ado About Nothing is very popular with audiences but somewhat less so with critics. Although it is conceded to be very witty, it is felt to be lacking in that profounder quibbling that characterizes Shakespeare's later work. In her book *Shakespeare's Wordplay*, M. M. Mahood gives a chapter to *The Winter's Tale* but not to *Much Ado About Nothing*. One may feel too that the play is less serious than Shakespeare's witty sonnets—for example, in its exploration of love. So far as the verse is concerned, it does not lead one to think of the play as a poem. It has a good deal of rather elementary rhetoric, as in Leonato's lamentations, and, although there are passages of charm and delicacy, perhaps no one would maintain that as poetry the writing ever equals the opening of *Twelfth Night* or Viola's "Make me a willow cabin." In fact, one of the most successful verse passages in the play—Hero's satire on the "lapwing" Beatrice—has the salience of wit rather than the ambience of poetry. The main plot of the play is certainly not the chief interest, and the central characters in this plot would never stimulate an A. C. Bradley. Moreover, the three main strands of action do not at first seem very well joined. The sudden appearance of Dogberry and his men in Act III, for example, comes as quite a jolt on the path of the action. The role of Margaret is mysterious, to say the least; only by straining can we think of her various activities as congruent.

William Empson once remarked that the greatness of English drama did not survive the double plot. Partially under Empson's influence, recent Shakespearian criticism is in general looking for Shakespeare's unities not in plot or character, or even characteristic action, but in theme. Actually, the theme of a play, if dramatically significant, is worked out in action, and conversely a particular action can be trans-

"The Role of Wit in Much Ado About Nothing*" by William G. McCollom. From* Shakespeare Quarterly, *XIX (New York: The Shakespeare Association of America, Inc., 1968), 165–74. Copyright © 1968 by The Shakespeare Association of America, Inc. Reprinted by permission of the editor.*

lated into theme. If you say, as does John Russell Brown,[1] that the theme of *Much Ado* is love's truth, the governing action (the activity guiding the characters) could be formulated as the search in love for the truth about love—though where this would leave Dogberry is a bit hard to say. In a keen study of the comedy, James Smith found pride or comic *hybris* the binding agent in an action presenting a shallow society whose superficiality is finally transcended by Benedick and Beatrice.[2] The analysis is illuminating, but I believe it pushes the comedy too far in the direction of satire and understates the role of wit, which in both its main senses drives the play.

During a performance of a Shakespearian comedy one sometimes notices that his neighbors are laughing at a line before the point has been made, or in ignorance of the exact meaning of the sentence, unless they have been studying footnotes. (This assumes that witticisms and jokes have exact meanings—not always a safe assumption.) One may feel a slightly superior sympathy for such an audience—they are so eager to enjoy what their piety has brought them to witness.

Yet this solicitude may be misplaced. For a witticism may be delightful and funny even if understood in a sense slightly different from that advanced by Kittredge or Dover Wilson. When Beatrice says that Benedick "wears his faith but as the fashion of his hat; it ever changes with the next block," [3] the audience laughs though it may not know whether "block" is a hat-block, a fashionable hat-shape, a block-head, or some combination of these. Secondly, if the actor has been advised to "throw away" the line as highly obscure and to create instead a visual and musical impression of wit, the audience can hardly be expected to laugh for the "right" reason. And finally, if Susanne Langer is right, the point of the line is not primary anyway; for Mrs. Langer advances the interesting idea that when an audience laughs, it does so not at a particular joke or witticism but at the play. In *Much Ado*, at any rate, wit is organic.

The wit of Shakespeare's play informs the words spoken by the characters, places the characters themselves as truly witty and intelligent, inappropriately facetious, or ingeniously witless, suggests the lines of action these characters will take, and, as intelligence, plays a fundamental role in the thematic action: the triumphing of true wit (or wise folly) in alliance with harmless folly over false or pretentious wisdom. I will further suggest that the comedy itself is a kind of witticism in the tripartite form often taken by the jests.

[1] *Shakespeare and His Comedies*, Second Edition (London, 1962), pp. 109–123.
[2] *"Much Ado About Nothing," Scrutiny*, XIII (1946), 242–257.
[3] *Much Ado About Nothing*, ed. David L. Stevenson, in *The Signet Classic Shakespeare* (New York, 1964), I. i. 71–73. This will be the edition cited.

As language, the wit has a variety of functions. From the first it creates the tone of "merry war" which will resound through so much of the comedy, though the timbre will change as the scenes or speakers change. The merry war is primarily between Benedick and Beatrice, but in the opening scenes Leonato, Don Pedro, Claudio, and even Hero participate in the skirmishing. Even before Don Pedro arrives with his party, we find Leonato experimenting with word-play. Hearing that Claudio's uncle has wept at the news of the young man's martial exploits, Leonato remarks: "a kind overflow of kindness. . . . How much better it is to weep at joy than to joy at weeping" (I. i. 27–28). It is as if he knew that some witty friends were coming to visit and he had better try out a pun and an antimetabole—a rhetorical figure popular in the earlier nineties. Since there has been no question of taking pleasure in tears, one tends to downgrade the speaker for this verbal flourish. But he may be more shrewd than this when, a bit later, he chides Beatrice for ridiculing Benedick: "Faith, niece, you tax Signior Benedick too much; but he'll be meet with you, I doubt it not" (ll. 44–45). Since "meet" and "mate" were pronounced alike, Leonato is not only referring to Benedick's powers of retaliation, but predicting the happy and voluble ending.

After establishing his fundamentally witty tone in the first three acts, Shakespeare almost destroys it in the church scene. But notice the language in which Claudio rejects Hero and Leonato responds to the scandal. There is the outburst of Claudio—

> O what men dare do! What men may do!
> What men daily do, not knowing what they do!   (IV. i. 18–19)

—a rhetorical display so hollow as to bring on this burlesque from Benedick: "How now? Interjections? Why then, some be of laughing, as, ah, ha, he!" As the scene progresses, Claudio's speeches rely more and more on the verbal tricks recorded in the rhetorical texts of the time. His half-ridiculous, half-pathetic pun "O Hero! What a Hero hadst thou been" is a parody of the wit crowding the early scenes. When he says:

> . . . fare thee well, most foul, most fair, farewell;
> Thou pure impiety and impious purity. . . .   (IV. i. 102–103)

the idiom is of the kind that Shakespeare will overtly ridicule at the turn of the century. Leonato's response to the rejection is equally conventional:

> But mine, and mine I loved, and mine I praised,
> And mine that I was proud on, mine so much
> That I myself was to myself not mine. . . .   (IV. i. 135–137)

The tone is precariously balanced between seriousness and levity. I believe that the scene has to be played for what it is worth and should not be deliberately distanced; otherwise the grief and anger of Beatrice will be unfounded; but if the dialogue is recognized as a distortion of wit, the scene becomes a grim sequel to the opening scenes and not an absolute break with them.

It is often difficult to separate style for tonal effect from style for characterization. But to put the matter in Renaissance terms, the decorum of the genre will sometimes take precedence over the decorum of the speaker. Critics like Stoll and Bradbrook have shown that the Elizabethans were frequently ready to drop consistency of characterization for tonal or other reasons. Margaret seems to illustrate the point. She is a witty lady-in-waiting, on excellent terms with both Hero and Beatrice, but the plot demands that she play her foolish part in the famous window scene that almost destroys Hero. After the rejection of her mistress, we see Margaret enjoying herself in a bawdy dialogue with Benedick, for all the world as if we were still in Act I. It is true that Hero has just been exonerated, but presumably Margaret does not yet know this. At the end of the preceding scene (v. i), Borachio has assured Leonato of Margaret's innocence of treachery to her mistress, but Leonato wants to know more. The men leave the stage, whereupon Benedick and Margaret enter for a set of wit. It is well played. But if we are trying to make sense of Margaret, we are puzzled. As she must be aware, her foolishness has been a main cause of all the distress, and she supposedly does not know of the happy solution brought about by Dogberry's men; if she does know, she also realizes that her role at the window is now revealed. Is she so indifferent to what has happened? Apparently we are not supposed to raise this question. Margaret asks Benedick if he will write a sonnet to her beauty.

> *Benedick.* In so high a style, Margaret, that no man living shall come over it; for in most comely truth thou deservest it.
> *Margaret.* To have no man come over me! Why, shall I always keep below-stairs? (6–10)

Margaret is here a representative of wit from the lady-in-waiting, and her quibble is related to her earlier wit but not to her earlier substantive behavior. Her wit at this moment is a bit crude. When Beatrice comes in a minute later, she will reveal a continuing concern for Hero along with a continuing mental agility. We can say that the two women represent two varieties of wit, though Beatrice is also clear as a character.

One has to distinguish between the seemingly ill-timed roguishness of Margaret and the really insensitive banter of the Prince and Claudio in Act v. Margaret makes no reference whatever to Hero, Leonato, or the painful episode of Act iv. But in v. i, after Leonato and his brother Antonio have quarreled with Claudio and Don Pedro over Hero and left the stage, Benedick enters, whereupon Claudio remarks, "We had liked to have had our two noses snapped off with two old men without teeth" (ll. 115–116). This is bad enough. Then, in view of Hero's supposed death, his cheery "What though care killed a cat" is one of his worst *gaffes*. When Benedick challenges his friend and tells him he has killed Hero, Claudio promises that in the duel he will "carve a capon." As the scene continues he and the Prince struggle to revive the tone of Act 1. As word-play, their language is much the same as ever, but neither Benedick nor the reader is in the mood for jocose references to "the old man's daughter," as if Hero were still happy. Stage directors and audiences seem ready to go along with the struggling wits at this point, but the reader's judgment is the right one: the scene makes a sardonic comment on the Prince and his young friend and gives supporting evidence of the ineptitude previously manifested. The wit *in this context* downgrades the two lords.

Apart from placing the characters, the play of wit indicates in advance the way the action will go. Where the repartee is not clearly out of place, the wittier speakers will prefigure in language the wit or intelligence of their acts. Benedick and Beatrice are the shrewdest in speech and with the Friar are the first to reject the rejection of Hero. What of Claudio's jests? At first they seem technically equal to Benedick's, but, on closer inspection, we notice that Claudio tends to repeat in somewhat different words the jests of the Prince. If Don Pedro heckles the amorous Benedick with "Nay, 'a rubs himself with civet. Can you smell him out by that?" Claudio will add, "That's as much as to say the sweet youth's in love" (iii. ii. 48–51). There may be a groundswell of laughter in the second line, but its point hardly differs from the other. If Don Pedro says that Beatrice has been ridiculing Benedick and then sighing for him, Claudio will chime in: "For the which she wept heartily and said she cared not" (v. i. 172–173). This echolalia illustrates the lack of independence which will cause him to swallow the slander of Don John and mirror the response made by the Prince. "O day untowardly turned!" says Don Pedro; and Claudio: "O mischief strangely thwarting!" (iii. ii. 127–128). Language is here the perfect expression of action, or rather of action descending toward comic automatism.

When Shakespeare was writing *Much Ado, wit* as mental agility or liveliness of fancy had rather recently come to supplement *wit* as intelligence. (A passage from Lyly is the first listing in *N.E.D.* of the newer use.[4]) Both senses occur frequently in the play, and there are examples of overlapping. It seems clear, for example, that in the following dialogue,

> *Dogberry.* . . . We are now to examination these men.
> *Verges.* And we must do it wisely.
> *Dogberry.* We will spare for no wit, I warrant you; here's that shall drive
>   some of them to a non-come.   (III. v. 57–60)

Dogberry is preening himself not only on his intelligence but on a handling of language so ingenious that it will drive the accused out of their minds. Benedick and Beatrice are witty and are described as witty and wise by their peers, and again both ideas are comprehended in the word "witty."

The word *wit* (or *witty*) occurs over twenty times, and one-third of these examples cluster in v. i, the scene in which Don Pedro and Claudio are flogging the dialogue. According to Benedick the wit does no more than amble in spite of the whip. As the scene progresses, one becomes weary of the verbal effort. After Benedick leaves, the Prince comments on his uncooperativeness: "What a pretty thing is man when he goes in his doublet and hose and leaves off his wit" (v. i. 198–199). Here the word suggests that for the idle nobility wit is a fashionable accessory you put on for lack of something else to do. In no other scene does this sub-sense (or Mood of *wit,* in Empson's terminology[5]) make itself felt.

In the drama, a particular witticism has three dimensions: the character's motivation for the speech, the technique, and the effect in context. A full criticism of a particular *mot* would have to consider all three. As Freud points out in his study of wit, a joke may be far more powerful than an examination of its technique would reveal: it may be poor in technique but strong in motive or "tendency." In a play, if a character's motive is strong, it may justify, in dramatic terms, what would be merely crude. Or if we share his animus, we will give way to hard laughter. In Act I Beatrice sometimes attacks Benedick in terms so unsubtle as to amaze—unless we realize that the insults express a half-conscious anger over his past treatment of her. At such moments we see the "wild" spirit of the "haggard of the rock" (III. i. 35–36), in Hero's phrase for her. The *effect* of a joke emerges in part from

---

[4] But William G. Crane shows that this sense goes back another forty years or more. See his *Wit and Rhetoric in the Renaissance* (New York, 1937), pp. 19–20.

[5] *The Structure of Complex Words* (London, 1951), pp. 17, 85.

motive and technique but may extend far beyond these. After Beatrice
has given a satiric picture of marriage, we have this:

> *Leonato.* Cousin, you apprehend passing shrewdly.
> *Beatrice.* I have a good eye, uncle; I can see a church by daylight.

(II. i. 80–82)

Leonato's speech is a mild rebuke but also an appreciation. The power
of her unforgettable reply is remarkable, considering the simplicity of
the technique, ironic understatement; but apart from the doubt
whether she is speaking modestly or proudly, the line looks back to her
own hardly masked fears of spinsterhood and forward to marriage in
general and Hero's ill-omened ceremony in particular, when Beatrice
will not only see the church but see better than most what is really
happening there.

The technique of wit in *Much Ado* may be classified under four
main heads:

(1) verbal identifications and contrasts including puns, quibbles, and
sharp antitheses;

(2) conceptual wit including allusive understatement and sophistical
logic;

(3) amusing flights of fancy;

(4) short parodies, burlesques, etc.

The first begins with the pun, as where it is said that Beatrice wrote
to Benedick and found them both "between the sheets." Claudio calls
this a "pretty jest," but Shakespeare uses the pun rather sparingly in
this play. Much more frequently are the quibbles wherein a speaker
deliberately mis-takes a word for his own purposes. Typically, a word
used metaphorically is suddenly given a literal sense: the Messenger
says that Benedick is not in Beatrice's "books," and she replies, "No.
And he were, I would burn my study" (I. i. 76). One is reminded of
Bergson's principle that it is comic to introduce the physical where
the spiritual is at issue. At the opposite extreme from the pun is the
sharp antithesis, as in Don John's assertion, "Though I cannot be said
to be a flattering honest man, it must not be denied but I am a plain-
dealing villain" (I. iii. 28–30). Here, of course, the wit includes
paradox.

I would suggest that, in comedy at least, the pun is a sign of harmony, the quibble or mis-taking is a ripple on the surface of social life,
and the antithesis an index of separation or selfishness. The pun is
obviously social and in comedy is seldom bitterly satiric. Even Claudio's
silly "what a Hero hadst thou been" is a sigh after vanished good

relations. Or one could cite Margaret's use of the pun as coquetry in her scene with Benedick. The quibble may be petty, but it is heavily dependent on what has just been said and may tacitly accept it. When Don Pedro declares that he will get Beatrice a husband, she replies that she would prefer one of his father's getting. The new meaning does not reject the old but merely improves it. The antithesis of Don John, on the other hand, flatly rejects the concept of "honest man," for like Goethe's Mephistopheles, John is the spirit that always denies. Since *Much Ado* is neither a jolly farce nor a morality play, it fittingly emphasizes mis-taking as opposed to puns and antitheses.

Freud's category of conceptual jokes or wit includes the play of ideas and playfully false logic.[6] The joke in Gogol, "Your cheating is excessive for an official of your rank," is conceptual, but it would become more abstract if transposed into the key of La Rochefoucauld as follows: "If a man appears honest, it is merely because his dishonesties are fitted to his position in life." Obviously the latter form is too abstract for *Much Ado,* though not for the tragedies. But in the mercurial world of *Much Ado,* Shakespeare infiltrates ideas less directly:

> *Don Pedro.* . . . I think this is your daughter.
> *Leonato.* Her mother hath many times told me so.
> *Benedick.* Were you in doubt, sir, that you asked her?
> *Leonato.* Signior Benedick, no; for then were you a child.   (I. i. 100–104)

Leonato's first pleasantry is standard social chit-chat, and no more critical than Prospero's "Thy mother was a piece of virtue, and / She said thou wast my daughter" (*Tmp.* I. i. 56–57). But Benedick's rude interruption, a quasi-quibble, pricks the complacencies of a cliché-ridden society. If his question is liberal, Leonato's reply is conservative: except for a few men like you, life in Messina is eminently respectable.

A good example of false logic in the service of true wit appears in Benedick's great soliloquy, which he speaks after hearing that Beatrice loves him. Faced with his own absolute opposition to marriage, he is capable of this: "Shall quips and sentences and these paper bullets of the brain awe a man from the career of his humor? No, the world must be peopled. When I said that I would die a bachelor, I did not think I should live till I were married" (II. iii. 236–240). Previously he had boasted that he would never decide to marry. Deserting that premise, he now pretends that the anticipated decision to marry can be taken as a mere occurrence happening to a thing innocent of choice. Involved in the complexity of the thought, however, is the speaker's awareness

---

[6] *Jokes and Their Relation to the Unconscious,* tr. James Strachey (New York, 1960), pp. 74 ff.

that to fall in love is to become a thing—an accident to which he grace-fully acknowledges himself liable. The soliloquy promotes Benedick from social critic to self-critic. He is now ready to appreciate the *maxime* of La Rochefoucauld: "C'est une grande folie de vouloir être sage tout seul." It is a crucial moment in the play.

At moments, Beatrice or Benedick will launch into an extended flight of fancy that moves distinctly away from its environment, par-ticularly because the play is dominated by prose. Benedick will describe a series of fantastic expeditions to escape Beatrice, or Beatrice will picture herself in a private harrowing of hell. Beatrice's comparison of wooing, wedding, and repentance to a Scotch jig, a measure, and a cinquepace is halfway between the conceptual wit just described and Shakespeare's more densely "tropical" style. Each of the dance steps is characterized as if it were a dramatic person, and all three encourage the actress to demonstrate. Like poor Yorick, Beatrice is a creature of "gambols," "of infinite jest, of most excellent fancy."

Outright burlesque or parody is used infrequently but significantly. When Beatrice asks Benedick if he will come to hear the news of Hero, he replies: "I will live in thy heart, die in thy lap, and be buried in thy eyes; and moreover, I will go with thee to thy uncle's" (v. ii. 100–102). This good-natured burlesque of the Petrarchan tradition affirms what we already knew, that Benedick will never make a con-ventional lover. Beatrice parodies Petrarchanism with deeper ironic effect:

> *Don Pedro.* Come, lady, come; you have lost the heart of Signior Benedick.
> *Beatrice.* Indeed, my lord, he lent it me awhile, and I gave him use for it, a double heart for his single one. . . . Marry, once before he won it of me with false dice; therefore your Grace may well say I have lost it.
>
> (II. i. 273–279)

In this moment she moves close to the atmosphere of the more somber Sonnets. The exploitation of the "usury" of love and of the dialectic of hearts recalls some of the opening Sonnets as well as the more intense poems to the Dark Lady.

Although the technique of wit in Beatrice's speech is good, it seems unimportant except as a revelation of motive or "tendency," in Freud's language. Nowhere else does Beatrice reveal so much of the reason underlying her war with Benedick, merry on the surface but now clearly shown to be serious underneath. If the seriousness were not there, she could scarcely keep her place as the wittiest of Shakespeare's characters. Beatrice had given her heart to Benedick as interest for his, but at the same time he received his own back again. But clearly there was another occasion when Beatrice felt she had been deceived into uncovering too much affection for him. In the nineteenth century such a motivation would bring on a suicide; in Shakespeare's play, it deepens the wit.

Seen as character, wit in *Much Ado* is awareness and the ability to act discerningly. As is already obvious, the awareness is largely the property of the talkative lovers. Such wit proves to be an Erasmian sensitivity to one's own folly. I have already referred to Benedick's increasing knowledge of his own limitations. Beatrice understands herself earlier. In the first scene she says to the Messenger: "[Benedick] set up his bills here in Messina and challenged Cupid at the flight; and my uncle's fool, reading the challenge, subscribed for Cupid and challenged him at the burbolt" (ll. 37–40). Dover Wilson thought she might be referring to a jester appearing in an earlier version of the play, but David Stevenson makes the excellent suggestion that the fool is Beatrice herself.[7] Other details strengthen the idea. Beatrice recalls the loss of her heart to Benedick. At another moment she names this heart a "poor fool." She fears that if she yields to Benedick, she will prove the "mother of fools." On which side of the family does she discern the folly? After entertaining the Prince with her merriment, she apologizes by saying: "I was born to speak all mirth and no matter." If she were a professional fool, she would not need to apologize.

Once Benedick and Beatrice have understood themselves, they are ready to act appropriately in the affair of Claudio and Hero. In the marriage scene Benedick immediately senses something deranged in Claudio's heroics, and when Hero faints under slanderous attacks, Beatrice immediately reveals her judgment: "Why, how now, cousin, wherefore sink ye down?" (l. 109). Whereas Leonato is completely convinced by the evidence, Beatrice is certain that Hero has been "belied."

Only after Beatrice has spoken out does the Friar join the defense of Hero. He has accurately read her character in her face.

> Call me a fool;
> Trust not my reading nor my observations,
> . . . .
> If this sweet lady lie not guiltless here. . . .
>
> (IV. i. 163–168)

At this moment he alone shares the wit of Benedick and Beatrice. Significantly, he is ready to be called a fool.

If wit marks the style and characterizes the dramatic persons in varying degrees, it is also the key to the "action"—taking this word in the

---

[7] "Introduction," *Much Ado About Nothing*, p. xxviii.

Stanislavskian sense as that focussed drive which unites all the larger and smaller activities of the play. From this point of view, the action of *Much Ado* is the struggle of true wit (or wise folly) in alliance with harmless folly against false wisdom. Don John, Borachio, Don Pedro, Claudio, and even Leonato represent in very different ways the false wisdom which deceives others or itself; Benedick, Beatrice, and the Friar embody the true wit which knows or learns humility. If we group the characters in this way, the conclusion of the play becomes more than the discovery of the truth about Hero followed by the double marriage but includes the triumph of true wit over false wisdom. The dominant tone of the play, however, finally softens the dichotomy I have suggested. The stupidities of the fine gentlemen are half-forgotten in the festive spirit of the close.

This interpretation of the basic action throws light on moments which might otherwise seem weakly articulated. One of these is the apparently rambling recital of Borachio to Conrade, as the Watch listen. These men, of course, stand for harmless folly as Borachio represents false wisdom. It was he who devised the entire plan to destroy Hero and who said, "My cunning shall not shame me" (II. ii. 55). His long digression under the penthouse emphasizes that although fashion—here equated with appearance—"is nothing to a man" III. iii. 119), young hotbloods will be deceived by it as Claudio had been deceived by Margaret's disguise. Borachio is shrewd enough to see the shallowness of the Claudios whom he can deceive but not wise enough to avoid boasting of his success.

Various references to fashion constitute a minor theme related to the theme of wisdom true and false. Preoccupation with fashion is a sign of immaturity or lack of wit. The unconverted Benedick is laughed at for being over-conscious of fashion, but in the climactic scene Beatrice flays the Claudios of society for their superficial and chic manners; "manhood," she says, "is melted into cursies, valor into compliment, and men are only turned into tongue, and trim ones too" (IV. i. 317–319). In his quarrel with Claudio, Antonio makes the same point. Antonio, who is often seen as a farcical dotard, strongly attacks "scambling, outfacing, fashionmonging boys" (v. i. 94). Properly read, the speech puts this old man on the side of wit as opposed to shallowness and takes its place in the not always obvious hierarchy of wisdom and folly.

A good play, like a good witticism, has a beginning, a middle, and an end. *Much Ado* is not only like wit; it can be seen as a witticism in tripartite form—the joke, of course, is on Claudio. In Freud's study of wit, there is a classification called "representation through the opposite." Like many other kinds of wit, this kind has three parts. It makes an assertion, seems to reaffirm it, but then denies it. A good example occurs in the following exchange from *Henry IV, Part I:*

> *Glend.* I can call spirits from the vasty deep.
> *Hot.* Why, so can I, or so can any man;
>   But will they come when you do call for them?   (III. i. 53–55)

Other witty exchanges can be reduced to the form: no, maybe, no—as here:

> *Leonato.* You will never run mad, niece.
> *Beatrice.* No, not till a hot January.   (I. i. 89–90)

Beatrice agrees, seems to have doubts, then agrees doubly. Many other examples of this tripartite form could be cited. I shall merely refer again to the comparison of wooing, wedding, and repentance to three dance steps. Here the witty sketch is a three-act play in little.

In the examples just given, the final proposition is not, of course, a simple denial or affirmation of the first. If the wit is to succeed, the climax must gain power through an obliquity which deceives expectation. The same method appears in some of the more ingenious Sonnets. Sonnet 139, "O call not me to justify the wrong," has the following structure:

(1) The lover asks the Dark Lady to refrain from wounding him by her straying glances.

(2) He argues that she is kind in looking aside since "her pretty looks have been mine enemies."

(3) He concludes that since her eyes have almost slain him already, they might as well kill him "outright" by looking straight at him. The conclusion returns to the opening, but with a crucial variation.

The beginning, middle, and end of *Much Ado* are not hard to name. The beginning is the successful wooing of the pure Hero. The middle is Claudio's conviction that she is impure: "Out on thee, seeming! I will write against it" (IV. i. 55). The end is the exoneration of Hero; but notice the words of Claudio:

> Sweet Hero, now thy image doth appear
> In the rare semblance that I loved at first.   (V. i. 252–253)

By this time the audience is convinced that the fashionmongering boy will never penetrate the reality lying beyond semblance. This is the joke on Claudio. He and his bride do not see the point, but the audience can hardly miss it.

As the play draws to its festive close, one may ask whether the friendship between Benedick and Claudio has essentially altered. The last scene would hardly be the place or time to say so. But in a final exchange, Benedick says: "For thy part, Claudio, I did think to have beaten thee; but in that thou art like to be my kinsman, live unbruised, and love my cousin." He must know that it will take some wit to do so.

"To Grace Harmony": Musical Design in
Much Ado About Nothing

by James J. Wey

Music, in its most general sense, occurs in *Much Ado* in the form of actual on-stage music[1] and allusions to musical properties in the dialogue. Actual music takes the more specific form of song, dance, and instrumental music alone or accompanying song or dance. In reading the play one is easily struck by the way in which the incidence of actual music reflects with unusual precision the main outline of action in the play. Some commentators, notably Long, have remarked upon this co-ordinate design of music and plot but only in a general way.[2] However, the co-ordination of musical reference with the thematic action in the play is much more precise and detailed than Long has observed it to be. And also, verbal allusions to music as well as the three forms of actual music are incorporated into the play's circle of meaning.

When we compare, therefore, the broad outline of actual music with

---

" 'To Grace Harmony': Musical Design in Much Ado About Nothing" by James J. Wey. From Boston University Studies in English, *IV* (Boston: Department of English, Graduate School, Boston University, 1960), *181–88*. [This article is a revised portion of a doctoral dissertation submitted to Catholic University of America and published in abstract form by The Catholic University Press, 1957.] Copyright © 1960 by Boston University. Reprinted by permission of the editor and author.

[1] This, of course, includes music heard off-stage for brief moments which, according to Kittredge's text and stage directions, occurs notably in I. ii. 1–26 and II. iii. 38–44. These and all ensuing textual references are to *The Complete Works of Shakespeare*, ed. George Lyman Kittredge (Boston: Ginn, 1936), pp. 159–192.

[2] For John H. Long the arrangement of music and song merely conforms to "the changing emotional appeals made during the course of the play" (*Shakespeare's Use of Music* [Gainesville: Univ. of Florida Press, 1955], pp. 120–121). For a discussion of some of the speculative assumptions about music which are part of the background to Shakespeare's use of music in *Much Ado* see the first half of John Hollander's *Musica Mundana and Twelfth Night* in *English Institute Essays: 1956* (New York: Columbia Univ. Press, 1957), pp. 55–82.

the general progression of the main plots in *Much Ado* we notice that actual music in some form occurs whenever there is harmony and happiness and love among the principal characters. Thus, as soon as the dramatic situation and chief characters are introduced in Act I, musicians enter and are put to work by Leonato (I. ii. 26). In the following act there is a dance in which the three sets of lovers, Hero and Don Pedro (representing Claudio), Beatrice and Benedick,[3] and Margaret and Balthasar[4] participate in a stately measure.

In II. iii music is heard within the house (lines 38–44) and Balthasar is then brought on stage (lines 44–92) to help in the plot which will eventually unite Benedick and Beatrice in love and matrimony. Now, however, as Don John's plot against Hero and Claudio unfolds, all forms of actual music which had hitherto accompanied love and harmony abruptly leave off, and the proposed serenade of Hero by Balthasar (II. iii. 86–89) is forgotten when love is replaced by hatred and harmony by disorder. Throughout the disruption of social peace and harmony in Act III there is no music of any kind. Even in the wedding scene at the church there is no music and certainly, if the actual music in the play had no dramatic significance and was merely ornament added to the play, there would have been some representation of music in this wedding scene.[5]

Only after Hero's name is cleared and disharmony begins to slacken is there music and song again. But it is, as yet, only Benedick's brief love lament (v. ii. 26–29) without instrumental accompaniment, and in the next scene Claudio's mournful song of repentance. It is only when complete harmony is restored with the union of all lovers, the apprehension of the villains, and the restoration of Hero from the "dead" that music reaches once again its earlier harmony and sprightliness. "Strike up, pipers," is Benedick's hearty close to the play, with stage directions for a dance.

Considered in this way, music seems to relate most obviously to the

---

[3] I am here following Kittredge's text closely. Charles T. Prouty, "A Lost Piece of Stage Business in *Much Ado About Nothing*" (*MLN*, LXV [1950], 207–208), believes it is Benedick who dances with Margaret in this scene and not Balthasar. If Prouty's text is accepted it would not substantially alter the significance of the dance as it is interpreted here.

[4] Borachio is Margaret's real lover, but Balthasar is obviously attentive to her also (II. i. 104–110).

[5] Other wedding scenes in Shakespeare include music or mention of music. See *The Taming of the Shrew*, III. ii. 185; and *Romeo and Juliet*, IV. iv. 22. The wedding in *All's Well That Ends Well* has no music, except what might be implied in the reference to a "solemn feast" (II. iii. 187). All marriages which have any preparation at all include music or at least mention of music, even when the wedding takes place off-stage as in *The Taming of the Shrew*.

Claudio-Hero plot. But one of the most interesting patterns in the play is the relation between the Claudio-Hero plot and the Benedick-Beatrice plot, especially in their common use of musical references. In this regard, in general, the Benedick-Beatrice plot maintains a kind of counterpoint to the Claudio-Hero plot.[6] In the first part of the play Benedick and Beatrice are busily rejecting love while Claudio is busily accepting it. With the slandering of Hero, Claudio rejects love while Benedick begins to accept it. Both plots conclude with the lovers reunited and wrongs righted.

In this way, for example, Benedick's rejection of love in the first half of the play is partly represented by his rejection of music; the absence of music in the middle of the play becomes symbolic of the wronging and slandering of Hero; and finally, the musical symbolism in both plots becomes one as the gradual revival of music symbolizes Benedick's acceptance of love and Claudio's repentance and final acceptance of Hero in marriage.

What makes this pattern most evident is not only the strategic occurrences of actual music which have been noted but also the presence of emphatic and controlled allusions to music which strongly associate song and music with love and courtship. Early in the play, for example, when Benedick thinks Don Pedro is going to woo Hero for himself instead of for Claudio as he had promised to do, he chides Don Pedro for stealing Claudio's "bird's nest." But, objects Don Pedro in the next line, continuing the bird image: "I will but teach them to sing and restore them to the owner" (II. i. 239–240). By "sing" in this passage, Don Pedro means "love"—(i.e., "I will but teach them to love")—and Benedick replies in the same figure: "If their singing answer your saying, by my faith you say honestly" (II. i. 241–242).

By this means song is blended metaphorically with love and this identification, together with other allusions, helps to identify love and music, and love and song so that the co-ordination of actual music, song, and dance with the main action of the play is made more conspicuous. This same association of music and love is made again in the dialogue which precedes Balthasar's song "Sigh No More, Ladies." Here the music of wooing is controverted into the wooing of music.

Enter *Balthasar* with *Music*
Pedro. Come Balthasar, we'll hear that song again.

---

[6] This counterpointing is carried out in many other ways as well. Both Benedick and Don John, for example, are subject to fits of melancholy and loss of appetite (II. i. 1 ff. and II. i. 152–156); both Benedick and Hero, like Don John, are called slanderers (II. i. 143; II. i. 245; and III. i. 84–86); and both the mock-slandering of Beatrice by Hero and the real slandering of Hero by Don John are spoken of as "poisoning" (III. i. 84–86; II. ii. 21; and v. i. 252).

> *Balth.* O' good my lord, tax not so bad a voice
> To slander music any more than once.
> *Pedro.* It is the witness still of excellency
> To put a strange face on his own perfection.
> I pray thee sing, and let me woo no more.   (II. iii. 44–50)

"Let me woo no more" is a variation on the idea of using music to woo such as occurred in the previous scene when Don Pedro used the occasion of the dance to woo Hero for Claudio. Like the metaphor of singing birds, this passage conflates again the theme of love and song.

Notice also in this passage how the theme of slander is woven into the conversation about song. Balthasar speaks of himself as being encouraged to "slander" music with his bad voice. The slandering of music (which is identified with love) thus makes another variation on the theme of slander, and in this indirect fashion—since later in the play love is actually slandered (in the person of Hero)—music is again associated with love.

Balthasar goes on in this same passage to further the analogy between himself as a singer and one who is being wooed.

> Because you talk of wooing, I will sing,
> Since many a wooer doth commence his suit
> To her he thinks not worthy, yet he wooes,
> Yet will he swear he loves.   (II. iii. 51–54)

In other words Balthasar is saying: Because you woo me to sing I can sing without pride because it is well known that men often woo women they know to be unworthy of their love. Therefore I can continue to believe myself an unworthy singer and yet sing because you woo me to it.

Still another musical allusion early in the play effects the same identification of music and love and makes an even more explicit reference to the three stages of the Claudio-Hero plot. This plot may be said to proceed from the wooing of Hero, to the wedding at which Hero is shamed, to Claudio's repentant song. These three stages of the plot—wooing, wedding, and repenting—are alluded to in an extended musical figure in Beatrice's advice to Hero as to how she should receive the anticipated proposal of marriage by the Prince.

> The fault will be in the music, cousin, if you be not wooed in good time.
> If the Prince be too important, tell him there is measure in everything,
> and so dance out the answer. For, hear me, Hero: wooing, wedding, and
> repenting is as a Scotch jig, a measure, and a cinquepace: the first suit is
> hot and hasty like a Scotch jig—and full as fantastical; the wedding,
> mannerly modest, as a measure, full of state and ancientry; and then

comes Repentance and with his bad legs falls into the cinquepace faster
and faster, till he sink into his grave.   (II. i. 72–83)

Here love, courtship, and marriage—the cycle of life with which the
play is chiefly occupied—are put into musical terms. "Wooed in good
time" means not only "wooed before too long" but "wooed according
to the musical measure." "Measure" means in this context not only
"moderation," in the sense that the Prince must not carry too far the
rule of marrying only in his own social class, but also "measure" in the
musical sense that he too must keep in step in the dance of love, "and
so dance out the answer." The comparisons between the three stages
of love and marriage and dance are so direct that they need no discussion, although the technical nature of these dances has been thoroughly discussed elsewhere.[7]

It is by such allusions to music as well as by virtue of the general
sequence of actual music that music takes on the special significance
it has. And once such significance is established it becomes the source
of further counterpointing of the two main plots. As we have seen,
music as a symbol of love parallels the Claudio-Hero plot. But at the
same time it is used as another way of displaying Benedick's rejection
of love.

Benedick continually speaks of love—and especially of his contempt for courtship and marriage—in musical terms. For example, he
shows in the same breath his contempt for lovers—and for ballad-
makers.

> Prove that ever I lose more blood with love than I will get again with
> drinking, pick out mine eyes with a ballad-maker's pen and hang me up
> at the door of a brothel house for the sign of blind Cupid.
>
> (I. i. 252–256)

And he describes Claudio's conversion to love as a change in musical tastes from the musical instruments of war befitting a man's interest to the weak and effeminate musical instruments of clowns and
fools.

> I have known when there was no music with him but the drum and the
> fife; and now had he rather hear the tabor and the pipe. (II. iii. 13–15)

Elsewhere Benedick's attitude toward love as reflected in musical
allusions is coupled with the cuckoldry motif which runs through the

---

[7] For descriptions of these dances see Gerda Prange, "Shakespeares Äusserungen
über die Tänze seiner Zeit," *Shakespeare Jahrbuch*, LXXXIX (1953), 132–161; also
Mabel Dolmetsch, *Dances of England and France* (London: Routledge and Kegan
Paul, 1949).

play.[8] In the first scene Benedick acknowledges his dependence on women for his birth and breeding, ". . . but that I will have a rechate winded in my forehead, or hang my bugle in an invisible baldrick, all women shall pardon me" (I. i. 242–245). The concealed pun on horn conflates the themes of cuckoldry and music and gives additional point to Benedick's later remarks about a "horn" in the next act when Don Pedro provides music as an appropriate prelude to the eavesdropping episode calculated to incite Benedick's love for Beatrice. Just before the song[9] by Balthasar, Benedick says to himself in an aside:

> Now divine air! Now is his soul ravish'd! Is it not strange that sheep's guts should hale souls out of men's bodies? Well, a horn for my money, when all is done. [*Balthasar sings.*]   (II. iii. 60–63)

Benedick here intends to show his contempt for song, especially for a love song, and the lute—another instrument of a lover's art. Benedick prefers the horn—the instrument associated with the pursuits of war and hunting. This is what the allusion to horn ostensibly means. But "horn" is also a symbol of cuckoldry which in this passage is the concealed pun foreshadowing with ironic overtones Benedick's final acceptance of marriage. Thus one innocent line anticipates his fall from single blessedness.

This scene is the turning point in Benedick's conversion to love. For not only does he make inadvertent allusion to his fate by reference to horn but he is apparently affected by Balthasar's carefully selected song and by the conversation which follows it. It is not long after this that he confesses his future wife must be, among other things, "an excellent musician" (II. iii. 36) and finally signals his complete capitulation to Cupid by singing a love song himself (v. ii. 26–29).

One again, however, music has application simultaneously to both plots. For Benedick's song occurs at the time when the slander against Hero has been found out and therefore it is part of the gradual return of music and song which slowly changes the mood of the play to its

---

[8] For other important allusions to cuckoldry which make Benedick's allusion to "horn" more conspicuous, see especially I. i. 202–204, I. i. 262–270, v. i. 183–186, v. iv. 43–45, and v. iv. 124–126.

[9] This song, like the remaining two songs in *Much Ado,* is admirably suited to the dramatic context. It is a love lament from the woman's point of view in which women deceived by men's inconstancy and indifference are advised to forget the man and be merry. This is good psychological softening of Benedick for it helps convince him that what he overhears his friends saying is true—that Beatrice's gay exterior is evidence that she inwardly sighs for him. In fact, Don Pedro's account in a later scene of Beatrice's state of mind closely echoes the sentiments suggested for the woman in the song (v. i. 168–176).

earlier gaiety. Benedick's song, therefore, is suitably sung without musical accompaniment, nor does he fully succeed in rhyming or singing it but rather breaks off with criticism of his own poor voice. In the scene which follows, Claudio delivers his penitential dirge over Hero's supposed grave—this time with sad and solemn words and music. In the next and final scene, with all evil absolved and love and harmony restored, the play turns full circle to its former lively music and dance.

Aside from these symbolic functions involving the lovers, music also is another device to set off the villains, especially Don John, from the other of the Prince's guests. Benedick, for example, implicitly accepts music even as he verbally rejects it and Don Pedro will teach the lovers to "sing," but Don John rejects music outright: "I am trusted with a muzzle and enfranchis'd with a clog; therefore I have decreed not to sing in my cage" (I. iii. 34–36). And later on in Act II there is an explicit stage direction suggesting that Don John and Borachio are to stand aside conspicuously and look on during the dance.[10] This emphatic musical characterization of a Don John who will neither sing nor dance is brought into sharper focus when we recall that on Hero's wedding eve it is Don John's slander which replaces the serenade which Balthasar had planned to sing beneath Hero's window (II. iii. 87–89).

In addition to all of this musical design there is one passage which suggests that Shakespeare intended a musical pun in the title of the play. It occurs fittingly enough at the conclusion of conversation in which, as we have seen earlier, Don Pedro "wooes" Balthasar to sing.

> *Pedro.*   Nay, pray thee come;
>   Or if thou wilt hold longer argument,
>   Do it in notes.
> *Balth.* Note this before my notes:
>   There's not a note of mine that's worth the noting.
> *Pedro.* Why, these are very crotchets that he speaks!
>   Notes notes, forsooth, and nothing!   (II. iii. 54–59)

The pronunciation of *nothing* as *noting* in Shakespeare's time seems well established and in view of the prominence of music in this play one might argue from the apparent pun in this passage that the title of the play involves a musical meaning among others.[11] Or at very least

---

[10] In Kittredge this reads: "Enter, [masked,] *Don Pedro, Claudio, Benedick,* and *Balthasar.* [With them enter Antonio, also masked. After them enter] *Don John* [and *Borachio* (without masks), who stand aside and look on during the dance]."

[11] See Kittredge's note in *Sixteen Plays of Shakespeare* (Boston: Ginn, 1946), p. 122n: "Thus Don Pedro repeats Balthasar's mock-modest remark and comments on it. 'Upon my word, he actually talks of "noting notes" and "noting"—and it all amounts to *nothing!* *Nothing* was almost or quite identical in pronunciation with

Don Pedro's musical quip suggests a key for interpreting the play, the spirit in which it should be taken. For music permeates the play. "How still the evening is" says Claudio, "as hushed on purpose to grace harmony." And more than the silence of the evening, everything in the play, in fact, is purposed to grace harmony. For just as the silence and moodiness of Don Juan (who is "out of measure sad" because "there is no measure in the occasion that breeds") only emphasize the bright and incessant conversation around him, so too the cacophony of slander and death rising momentarily above the mock discords of Benedick and Beatrice only emphasizes the happy return of love, life —and music.

---

noting. It rhymes with *doting* in the 20th Sonnet." More recently Helge Kökeritz (*Shakespeare's Pronunciation*, New Haven: Yale Univ. Press, 1953, p. 132) has rejected an older view that "noting" in Elizabethan usage can mean "eavesdropping" but he adds that according to meanings listed in *OED* it could mean "to brand with disgrace, to stigmatize." This passage of Don Pedro's could therefore refer to the theme of slander and the title of the play would involve a triple pun of "noting" in a musical sense, in the sense of slander, and in the usual sense of "nothing." For a discussion of other uses of this same pun elsewhere in Shakespeare, see Paul A. Jorgensen, "Much Ado About Nothing," in *Shakespeare Quarterly*, v (1954), 287–295.

Music for *Much Ado About Nothing*

by Virgil Thomson

Adding music to Shakespeare's plays is for the composer a discipline of modesty. Some music is invariably required; but this has been so carefully limited by the poet, boxed in to the play's bare needs, that one comes out of the experience convinced that the Bard was wary of all music's disruptive dangers and ever so careful lest musicians, a powerful and privileged group in Shakespeare's England, steal the show from poetry.

The tragedies mostly ask for one song, rarely more. Occasionally it is possible to add at the end, as in *Hamlet,* a funeral march. The rest of the music can be done with trumpets and drummers, two of each. They announce royal exits and entrances, indicate from offstage the advance and retreat of soldiers, even evoke (by an antiphony of pitches and motives) armies in combat, triumphs, and defeats. The drummers can also, on cymbal, bass drum, thunder drum, tamtam, bells of all sizes, wind machines and other sound effects, produce the storms, fogs, and other species of foul weather that play so important a part in the tragedies.

It is usually better to produce these effects through musicians rather than through stage hands, to score them and conduct them, to control them in timing and volume through a series of electric light cues coming directly from the stage manager. (No one can hear word cues while playing percussion.) Only in this way, and after much rehearsal, can they be used to build an actor's vocal resonance acoustically, like a good musical accompaniment, rather than compete with it, fight against him, destroy his finest flights.

The comedies require few battles and less weather, allow for a bit more straight music. There may be an extra song, sometimes a moment of dancing, the evocation of a balmy night, a brief wedding, and, instead of the funeral march, with luck a gracious musical ending not

"*Music for* Much Ado About Nothing" *by Virgil Thomson. From* Theatre Arts, *XLIII (June, 1959), 14–19. Copyright © 1960 by Virgil Thomson. Reprinted by permission of the author.*

unlike our modern waltz finales. In general, however, the music cues are brief and, as in the tragedies, limited to the minimum needed for creating an atmosphere. Nowhere in Shakespeare, save in *Henry VIII,* which is not wholly from his hand, will the composer be allowed to create the massive ecclesiastical ceremonies, grand marches, ballets, balls, and dream sequences that abound in Marlowe and Webster. Everywhere the music must be straightforward, speak quickly, take no time at all out of the play's dramatic pacing.

Under modern union rules and in view of heavy costs, the musicians may be as few as four in number. The choice of instrumentalists for meeting the play's musical requirements with a minimum of aural monotony is a matter for much care. Gone are the days when Mendelssohn could use a fully balanced orchestra for *A Midsummer Night's Dream,* filling the theatre with symphonic sound and holding up the play for the musical working out of themes, the injection of intermezzos and of expansive "numbers." Today's stage music must represent economy in every domain.

Stylistically speaking, music for a spoken play must always be subservient to the visual element, as well as to the verbal. A production, first conceived by the director, must be sketched for sets and costumes before the composer of incidental music can effectively plan his contribution. A sumptuous visual style needs richness in the auditory too, while a meager, or bare-stage production invites music of minimal luxuriance. In every case, music is an extension of the decorative scheme.

Also there is a matter of time and place. Many a director and his stage designer find it advantageous to take a script out of its period. Sometimes the script itself gives such a choice. *Hamlet,* for instance, can with equal ease be put into a medieval Denmark, into a Renaissance Denmark (contemporary with Shakespeare), or into modern dress. Music should accentuate, underline the chosen background. Consequently, it should not, for the best result, be even thought about by the composer before the period references of the production have been determined. Suggesting through the musical style of Richard Wagner or of Tchaikowsky the plush comforts of Queen Victoria's time (which was theirs) or using neat Mozartian turns and balances, however appropriate these might be in a setting of eighteenth-century court life, will not be of much help to *King John* or *The Merchant of Venice.* Stage music need not be historically authentic; but like the settings and the costumes it should evoke whatever time and place the director has chosen for his production.

The performances of *Much Ado About Nothing* that were first produced at the American Shakespeare Festival Theatre in Stratford, Con-

necticut, during the summer of 1957 represented in their visual and
musical design exactly the sort of director's choice that I am describ-
ing. This choice has seemed arbitrary to many. Taken originally by
the co-directors, John Houseman and Jack Landau, the decision was
accepted with delight by Katherine Hepburn and Alfred Drake, the
Beatrice and Benedick, as well as by Reuben Ter-Aratunian, who designed the sets and clothes, and by the present writer, who wrote music
for the production. It seemed to us all, and still does, I think, imaginative and wise, a contribution, even, to the play and to its comprehen-
sibility in our time.

This decision, indeed a radical one, was to set the play somewhere
in northern Mexico, in what is now Texas or southern California, in
the middle of the nineteenth century. Shakespeare had put it in a Sicily of his own time, or a little earlier, under the Spanish occupation.
Now this locale is not a vague one like "the seacoast of Bohemia" in
A Winter's Tale or the Illyria of *Twelfth Night* or the even vaguer
Vienna of *Measure for Measure,* slight masks all of them for English
country life or for London. The people in *Much Ado* are as charac-
teristically Spanish as those in *Othello* and *The Merchant* are Vene-
tian. They are proud, simple, sincere, energetic, and deeply sentimental. Even the villain is proud of his villainy and quite frank about it.
None of them has the malice, snobbery or vainglory, the need of being
two-faced about things, the divided mind or the indirection that
abound in Shakespeare's other plays. Everybody is childlike, surely
Catholic, and deeply, tenderly, permanently oriented toward affection.
Also there is no war between the younger and the older generations.

Since Shakespeare says these people are Spanish colonials and since
their characteristics are in fact recognizable today as those of Spanish
colonials, changing their nationality would be a mistake. Surely the
young Claudio who, believing he has caused the death of his beloved
by doubting her virtue, accepts as a penance for his sin to marry sight
unseen a cousin (non-existent) invented by his fiancée's father is none
of Shakespeare's English young men. Only as a Spaniard, believing in
sin and its consequences, fatalistic, loyal to the results of his own actions, can he be believed and sympathized with. No matter how the
setting of it may be altered, the play must remain a picture of life
among well-to-do Spanish landed gentry.

The Sicilian background is less demanding, not demanding at all in
fact. The monuments of that isle, its history and its geography, save
for some casual mention of Messina, play no part. Using Sicilian
landscape or architecture, Hispano-Sicilian clothes, especially from
three or more centuries back, would tend to confuse rather than to

clarify the play's understanding today. And transporting it to Shake-speare's England, always a possible choice, would not be a happy one, so thoroughly un-English is the love story.

And so, as a result of meditations and of searches after other alter-natives, the idea was conceived that a rich ranch house in Spanish North America would be just the place for Hero and her father, for Claudio and his young friends just back from brilliant conduct in some military action. Beatrice and Benedick, of course, fit into any time or place that offers a background of mating and matchmaking. The whole play, it was agreed, could be brought to life in such a set-ting with a maximum of recognition on the part of twentieth-century Americans and a minimum of violation to the poet's meaning.

Now the music cues in such a setting are no different in placement or in length from what they would have been had the play been left in Sicily. But their character, their whole style and texture, were of necessity determined by the adopted time and place. Like the scenery and the costumes, the easy movements and the formal dances, they had to evoke nineteenth-century Spanish America.

With this in mind I went to the New York Public Library and ex-amined all the collections of folklore and old popular tunes from northern Mexico and the American southwest. I copied out everything that I thought might possibly be useful. Then I consulted with my di-rectors to determine the spots where music might be needed, desired, possible to introduce, and about the number of musicians the produc-tion budget would allow.

As it turned out, I got nine instrumentalists; I chose a flute (dou-bling on piccolo), a clarinet, one percussion player (who could also play bells, or *glockenspiel*), two trumpeters, one viola, one cello, and a man who could play both guitar and double-bass. In addition, the management provided just for singing the songs, the famous counter-tenor Russell Oberlin (dressed as a peon of the ranch). There was also a conductor, not really necessary; but since he was attached to the the-atre that season, and could also play the piano, he was a useful luxury. I arranged placement for my musicians in a covered pit, where they could be heard but not seen; and the electrician placed microphones and loud speakers so as to amplify my string section without the pres-ence of amplified sound being detectable to the ear.

I shall not give here the complete list of music cues, because we did not follow the script of the play exactly. Virtually no director does nowadays follow a Shakespeare script as it appears in a scholarly edi-tion. He makes cuts for brevity and cuts for modern pacing, which permits less rhetorical dalliance than audiences liked in earlier times.

He transposes the order of scenes for clarity and for scenic convenience. He allows only one or two intermissions for smoking and refreshment. He may add scenes of pantomime and dancing, if the story invites them. He may even interpolate an extra song from some other Shakespeare play, if he has a charming singer, or omit those already there, if he does not. In short he cuts and re-shapes the play to a modern audience's taste for speed and a clean trajectory. He does not, of course, re-write the prose or poetry. What he does is comparable to the cuts, transpositions, instrumental substitutions, and adjustments of orchestral balances that any conductor makes for the performance today of an oratorio by Bach or Handel. Such changes are every bit as legitimate as using women instead of the boys that Shakespeare's time required in the female roles. But they do make every director's production a "version," and they do determine for that version the placement of the music cues.

In the Houseman-Landau version of *Much Ado,* the soldiers returned from war by marching down a main aisle of the theatre and on to the stage, led by fife and drum playing a quickstep. For this I used the most commonplace tune possible, the familiar "Jarobe Tapatio," generally known as "Mexican Scarf Dance," and speeded it up for excitement. I later used "La Golondrina" too, this timed as a habanera. These tunes, plus another almost equally familiar, reappeared in different instrumentations several times, as if they were familiar banalities of the time and place (which, indeed, they were). They helped to establish a folksy and easy-going Spanish-American ambience. They also served to set off by contrast the more refined musical backgrounds of the ball scene, the wedding, the tomb scene, and the happy ending.

For these latter I found authentic bits from the old Southwest, took two dance tunes from Spain, and composed a waltz. I needed a waltz for romantic feeling. But our Spanish-Americans had no waltzes of their own. They used (bought copies of it by the thousands) a German waltz called "Over the Waves" (to them, "Sobre las Olas"). This did not seem right to me to use; it is too familiar to be effective save in ironic or caricatured version. And I needed a waltz with some lift in it. So I wrote one that might have come from anywhere, that evoked the romantic feeling that upper-class people everywhere in the nineteenth century associated with three-four time.

The famous song beginning "Sigh no more, Ladies" I composed in two versions and used in two places. One version was in espagnoloid Scottish style, with syncopations derived naturally from the word accents and a vocal flourish at the end. The other was a fandango in which the words clicked like castanets. This latter so deeply offended the taste of my singer, an expert of Elizabethan songs and madrigals, that eventually it was removed from the production.

It so happened in this production that almost every cue was "source" music, a situation ever to be wished for. "Source" music means, in show business, music which can be supposed to come from some source, on or offstage, that is part of the play. It includes the trumpets that announce a king, the drums of the military, anybody singing a song, music that might be part of a household, such as dance music, dinner music, anything present or overheard that offers a realistic excuse for being present or overheard. This kind of music is truly "incidental," though it may be as necessary as any property sword or goblet. And it is the most expressive of all play music, because, its presence being dramatically explained, it can underline moments of tenderness, produce suspense, emphasize climaxes, aid the play through music's great power of producing emotion, without the audience experiencing the shock that comes from a breaking of the dramatic illusion.

More dangerous, but sometimes indispensable for creating a mood, is music that is purely atmospheric, an auditory addition to the scenery. Forests, moonlit gardens, shooting stars (I once produced this last effect through one chord on a celesta), all those states of nature that need to be felt as palpable but that cannot be rendered delicately enough by scenic or lighting effects, can be, must be, essayed through music. Here the music has a poetic, not a realistic, source. It is the voice of Nature.

Music can also speak with the voice of Memory. Soliloquies, confessions of yearning, recalls of innocence and childhood, can without embarrassment to the listener be accompanied by soft music of specifically evocative character. Shakespeare, we are told, used for this purpose recorders, a species of primitive flute. Intermezzos and introductions, when not truly "source" music, can also recall, announce, or anticipate without stepping out of the play. Overtures, of course, can only announce and consequently are best when justifiable as "source" music. Anything else becomes the voice of the manager, the "barker" at a sideshow, and is ill-suited to the poetic theatre. An overture that is like a preview of the scenery can be very effective, however.

Beyond these conventions I know no musical usages that are appropriate to the poetic stage. "Source" music is always the best, but the voices of Nature and Memory need not be excluded. They must only be handled with tact. Occasionally a brief sound effect can be good too, if introduced and ended before the audience has time to realize its instrumental source. This is like physical pain made audible. A tamtam roll *ppp poco crescendo* as Othello's epilepsy comes on him, a disembodied electronic whine as Banquo's ghost appears to Macbeth,

such effects are in the best Elizabethan tradition. If they come off, they
are very powerful. But if anyone in the audience laughs, they must be
abandoned.

Among the voices I do not consider it proper to represent through
stage music are those of the author, who has chosen language as his
idiom and should not step out of that convention, the voice of the
producer, who is no part of the play, and those of the director, designer, costumer, and composer. These collaborators are speaking, we
may presume, for the author by helping the play to speak for him.
Their self-exploitation is as offensive as that of an actor who insists on
playing some publicized or imaginary image of himself instead of the
role we have come to see him do. Stage music, to serve well in performance, must be objective, modest, and as loyal a collaborator of
the director's and the designer's plan as of the author's play. That way
lie all the possibilities of boldness and of daring. And that way too,
so far as Shakespeare's plays allow music at all, lies the only possibility
there is for a composer to serve, as a worker, the great Shakespearean
texts.

In our Spanish-American *Much Ado,* there was a short overture, a
habanera, to set the locale. Its "source" in the play's reality might be
considered to be a café or dance hall outside the gates of the ranch,
where our first scene was placed. The soldiers returned and disbanded
here to a *paso doble,* or quickstep, led by musicians in costume.

There was reception music, Hispanic, for the opening of our second
scene. From time to time a sentimental Spanish melody, one of those
heard elsewhere in the play as café or dance music, was overheard on
a guitar beneath some love scene or soliloquy. And there was formal
dancing to music during the masquerade. Also, waltz music for the
grand march out to supper. Balthasar sang his song to guitar accompaniment. And we ended our Act I on Beatrice's soliloquy with waltz
music under it, like a memory, crescendo.

Act II began with café music, and the Dogberry scenes of municipal
justice were framed in similar material. Dressing the bride was pre-
ceded and in part accompanied by an *alborado,* a salute to the marriage morn. Later there were wedding bells and wedding music. Then
more café music, dance tunes on a tinny piano, to mark the police
action as taking place among persons of banal taste, whose lives were
drenched (literally) in hand-played bar ballads, as ours are in mecha-
nized musical banalities.

Exactly such a tune opened our Act III, and comic drums accompanied a change of guard at the police station. Benedick's song, brief,
amateurish, and absurd, was not accompanied. Then came funeral
music for the tomb scene and for the song, "Pardon, Goddess of the

Night," followed by wedding music again, by dance music, and at the end by a fanfare and general waltzing.

Transitions from one scene to another were usually covered by music, either in continuation of the preceding mood or in anticipation of what was to follow. We put as much music into the play as we could, endeavoring thus to keep the comic and the sentimental tones constantly present and warm. We hoped also that the abundant presence of music on two levels of taste—the vulgar and the genteel—would help us to underline the basic premise of the production, which was that the story takes place on a vast estate with many sorts of people around and many kinds of life going on. It is all a shade provincial; but there are great hospitality, many arrivals and departures, justice administered casually, military forays, cooks, chambermaids, peons, and musicians constantly available for producing a ball, a wedding, a funeral, anything needed in the life of this rich, easy-going, isolated Spanish family.

My music cues would not be applicable stylistically to a production differently conceived. I have described them here for what they were, an auditory extension of the scenery and costumes. These latter illustrated visually the directors' conception of the play. This conception was based on what the directors and their collaborating artists considered to be the nature of the play's people and message. Theirs is not the only legitimate view. But once decided on and accepted as a basis for production, it determined the character of the scenery and of the music. As for the amounts of music used and its exact placement, we were obliged, in spite of all our efforts toward abundance, to limit our largesse. The controls that Shakespeare once and for all built into this script, as he did into every other, cannot be broken through without doing violence to the play. To make an opera or an operetta out of a Shakespeare play requires abandoning most of the Shakespeare in it.

PART TWO

View Points

Harold C. Goddard

If I draw a circle on the sand or on a piece of paper, instantly the spatial universe is divided into two parts, the finite portion within the circle (or the sphere if we think of it in three dimensions) and the infinite remainder outside of it. Actuality and possibility have a similar relation. Actuality is what is within the circle. However immense it be conceived to be, beyond it extends not merely the infinite but the infinitely infinite realm of what might have been but was not, of what may be but is not. In this realm are all the deeds that were not done when the other choice was made, all the roads that were not traveled when the other fork was taken, all the life that did not come into existence when its seeds failed to germinate. And in it no less is all that still may be: all the possible combinations of chemical elements that have never been made, all the music that is still uncomposed, all the babies that have not yet been born. This is the realm of NOTHING. In one sense it has no existence. In another, existence is nothing without it. For out of it ghosts are perpetually being summoned by our hopes and fears—which are themselves made of nothing—to be incorporated into the world of FACT. The interflow and union between these two realms is the type and father of all the alchemies and chemistries. "He who has never hoped shall never receive what he has never hoped for." Thus Heraclitus packs it into a sentence. *Ex nihilo nihil fit* is but a half-truth. For out of something nothing new ever came without the aid of "nothing" in this high potential sense. Nothing is thus practically a synonym for creativity. It is that realm of pure possibility that alone makes freedom possible. It is one of the two constituents of the imagination, the other being fact.

Shakespeare delighted in using the word "nothing" in this high metaphysical sense. It is easy to see why any artist might.

* * *

Now if our eyes are not so dazzled by Beatrice and Benedick and

From "Much Ado About Nothing" *by Harold C. Goddard. From* The Meaning of Shakespeare, *2 vols. (Chicago: University of Chicago Press, 1951), I, 271–72, 273, 275. Copyright © 1951 by the University of Chicago. Reprinted by permission of the publisher.*

the glitter of their wit, or our risibilities so tickled by Dogberry and
his companions, that we cannot attend to the play as a whole, we shall
see that it is dedicated to this idea of Nothing. It is full of lies, deceptions (innocent and not so innocent), and imagination, and these
things grade into one another as imperceptibly as darkness does into
light. Yet, notwithstanding that fact, the extremes—namely, lies and
imagination—are seen to be as opposite as night and noon.

         *   *   *

Much Ado About Nothing is saturated with this idea of the power
of Nothing (of the creative ingredient of the imagination, that is) to
alter the nature of things for good or ill, for, as Shakespeare's History
Plays so abundantly show, fear and hate, as well as faith and love,
have the capacity to attract facts to them and so, temporarily at least,
to confirm their own hypotheses. But the changes of fear and hate effect
are destructive and pointed in the direction of chaos, whereas imagination integrates, makes for synthesis and reconciliation of clashing
interests. The play is full of phrases that imply this fluidity of facts,
their willingness to flow for good or evil into any mold the human
mind makes for their reception. Antonio brings news to Leonato. "Are
they good?" asks the latter. "As the event stamps them," the former
replies. "You have of late stood out against your brother," says Conrade to Don John, "and he hath ta'en you newly into his grace; where
it is impossible you should take true root but by the fair weather that
you make yourself: it is needful that you frame the season for your
own harvest." The children of this world, as Jesus divined, are often
wiser in these matters than the children of light. But not so in the
case of Friar Francis. "Die to live," is his advice to Hero, which is only
a more succinct summary of his prophecy of the effect on Claudio of
the "nothing" of Hero's death:

>   for it so falls out
> That what we have we prize not to the worth
> Whiles we enjoy it, but being lack'd and lost,
> Why, then we rack the value, then we find
> The virtue that possession would not show us
> Whiles it was ours. So will it fare with Claudio:
> When he shall hear she died upon his words,
> The idea of her life shall sweetly creep
> Into his study of imagination,
> And every lovely organ of her life
> Shall come apparell'd in more precious habit,
> More moving-delicate, and full of life

Into the eye and prospect of his soul,
Than when she liv'd indeed.

Dorothy C. Hockey

Richard Grant White's suggestion, made as long ago as 1857, that
nothing and *noting* constituted an Elizabethan pun[1] has recently been
seconded in Kökeritz's valuable study of Elizabethan pronunciation.[2]
Kökeritz, however, considers it "unlikely" that *noting* is used to mean
eavesdropping, as White had further suggested. In the words of Professor Jorgensen, "rejection" of this latter portion of the early critic's
theory is "implicit" in the "almost perfect editorial silence" that followed it.[3] Commentators on the play have been content to point to
the repeated use of eavesdropping and leave the matter there. But
noting in a sense understood today as well as in Shakespeare's day is,
I believe, the key to the play's thematic unity—*noting* meaning to
observe or, as Schmidt cites it in his *Lexicon,* "to attend to, to observe" (II, 780–781). We note a situation; we take note of a situation
—we see and hear, then judge and act accordingly. *Much Ado* is a
comedy of *mis*-noting in this common sense. Eavesdropping, then, becomes just one kind of observation. Throughout the play every character is required to observe and judge, and almost every character
judges poorly. Deception plays a part in these misjudgments, as Professor Prouty has pointed out,[4] but much more pervasive a force is a
common human frailty—the inability to observe, judge, and act sensi-
bly. The play, then, is a dramatization of mis-noting—a sort of dramatized, rather than verbal, pun.

*   *   *

Shakespeare signals his purpose in another way, too, the verbal pun.
In his opening scene when Claudio asks his friend, "Benedick, dids't

---

From *"Notes Notes, Forsooth . . ."* by Dorothy C. Hockey. From Shakespeare
Quarterly, *VIII (New York: The Shakespeare Association of America, Inc., 1957),*
353–54, 355. Copyright © *1957 by The Shakespeare Association of America, Inc.*
Reprinted by permission of the editor.

[1] *Much Ado About Nothing,* New Variorum edition (Philadelphia, 1899), pp. 6–7.
[2] Helge Kökeritz, *Shakespeare's Pronunciation* (New Haven, 1953), p. 132.
[3] Paul A. Jorgensen, "Much Ado About *Nothing," SQ,* V, 294.
[4] Charles T. Prouty, *The Sources of Much Ado About Nothing* (New Haven,
1950), pp. 51–52. This study reveals that the eavesdropping is not to be found in the
sources.

thou note the daughter of Signior Leonato?" (l. 164), he prompts this reply, "I noted her not, but I look'd on her" (l. 165). A more striking use of the verbal pun occurs, of course, in the dialogue between Balthazar, the singer, and Don Pedro in II. iii. To end Balthazar's polite noises about his poor voice, Don Pedro bids him

> Nay, pray thee come:
> Or if thou wilt hold longer argument,
> Do it in notes.
> *Balth.* Note this before my notes:
> There's not a note of mine that's worth the noting.
> *Pedro.* Why, these are very crotchets that he speaks!
> Notes notes, forsooth, and nothing!   (II. iii. 54–60)

The textual difficulty of the last word is of little consequence at this moment. Whether Shakespeare wrote *nothing* or *noting*, the entire passage emphasizes the word *note* unquestionably, punning on the musical term and the idea of observing or heeding. The placing of this dialogue is interesting, too, for Shakespeare chooses the moment of Benedick's gulling for thus calling our attention once more to his thematic device of noting.

A. P. Rossiter

Without striving to make too much of it, the dance in II. i. is beautifully apposite. The couples walk their round, two by two, all masked; and all are using words to back the disguise of false faces with trivial deceit. The play-acted defamation of Hero, by means of a false dress on the wrong woman and names used falsely, is exactly parallel. In both, the truth is *behind* the looks and words. The *bal masqué* is only a game of seeming; yet it is a most apt symbol of the whole. The vizor is half deceit, half no deceit: you can never be sure. Believe it, and you make ado about what is nothing. And in the social order and shared delight of the dance—all moving to the controlling rhythm, in their appointed patterns—there is too the emblem of the harmony in which all will conclude: as the play does, with another dance, all the vizors laid aside. The real play is not ended with "Strike up, pipers." The very movement of II. i., where all the main misapprehensions started, is

From Angel with Horns *by A. P. Rossiter, edited by Graham Storey (London: Longmans, Green & Co. Ltd., New York: Theatre Arts Books, 1961), pp. 75–76. Reprinted by permission of the publishers.*

repeated and completed; and even the professed misogamists are dancing to the same tune. It is as neat and pretty as "Sigh no more, ladies,
sigh no more."

W. H. Auden

Much Ado About Nothing
Act ii, Scene 3.
Song. Sigh no more, ladies.
Audience. Don Pedro, Claudio, and Benedick (in hiding).

In the two preceding scenes we have learned of two plots, Don Pedro's plot to make Benedick fall in love with Beatrice, and Don John's
plot to make Claudio believe that Hero, his wife-to-be, is unchaste.
Since this is a comedy, we, the audience, know that all will come right
in the end, that Beatrice and Benedick, Claudio and Hero will get
happily married.

The two plots of which we have just learned, therefore, arouse two
different kinds of suspense. If the plot against Benedick succeeds, we
are one step nearer the goal; if the plot against Claudio succeeds, we
are one step back.

At this point, between their planning and their execution, action
is suspended, and we and the characters are made to listen to a song.

The scene opens with Benedick laughing at the thought of the love-
sick Claudio and congratulating himself on being heart-whole, and
he expresses their contrasted states in musical imagery.

> I have known him when there was no music in him, but the drum and
> the fife; and now had he rather hear the tabor and the pipe. . . . Is it
> not strange that sheeps' guts should hale souls out of men's bodies?—
> Well, a horn for my money when all's done.

We, of course, know that Benedick is not as heart-whole as he is trying
to pretend. Beatrice and Benedick resist each other because, being
both proud and intelligent, they do not wish to be the helpless slaves
of emotion or, worse, to become what they have often observed in oth-

---

From "Music in Shakespeare" by W. H. Auden. From The Dyer's Hand and
Other Essays *(London: Faber & Faber Ltd., New York: Random House, Inc., 1962),
pp. 516–18. Copyright © 1957 by W. H. Auden. Reprinted by permission of the
publishers.*

ers, the victims of an imaginary passion. Yet whatever he may say against music, Benedick does not go away, but stays and listens.

Claudio, for his part, wishes to hear music because he is in a dreamy, lovesick state, and one can guess that his *petit roman* as he listens will be of himself as the ever-faithful swain, so that he will not notice that the mood and words of the song are in complete contrast to his daydream. For the song is actually about the irresponsibility of men and the folly of women taking them seriously, and recommends as an antidote good humor and common sense. If one imagines these sentiments being the expression of a character, the only character they suit is Beatrice.

> She is never sad but when she sleeps; and not even sad then; for I have heard my daughter say, she hath often dream'd of happiness and waked herself with laughing. She cannot endure hear tell of a husband. Leonato by no means: she mocks all her wooers out of suit.

I do not think it too far-fetched to imagine that the song arouses in Benedick's mind an image of Beatrice, the tenderness of which alarms him. The violence of his comment when the song is over is suspicious:

> I pray God, his bad voice bode no mischief! I had as lief have heard the night-raven, come what plague could have come after it.

And, of course, there *is* mischief brewing. Almost immediately he overhears the planned conversation of Claudio and Don Pedro, and it has its intended effect. The song may not have compelled his capitulation, but it has certainly softened him up.

More mischief comes to Claudio who, two scenes later, shows himself all too willing to believe Don John's slander before he has been shown even false evidence, and declares that, if it should prove true, he will shame Hero in public. Had his love for Hero been all he imagined it to be, he would have laughed in Don John's face and believed Hero's assertion of her innocence, despite apparent evidence to the contrary, as immediately as her cousin does. He falls into the trap set for him because as yet he is less a lover than a man in love with love. Hero is as yet more an image in his own mind than a real person, and such images are susceptible to every suggestion.

For Claudio, the song marks the moment when his pleasant illusions about himself as a lover are at their highest. Before he can really listen to music he must be cured of imaginary listening, and the cure lies through the disharmonious experiences of passion and guilt.

James A. S. McPeek

The imagery of *Much Ado* can be separated from the action only by an act of violence, so much is it a part of the texture of the play. This is particularly true of the predominant imagery, that of fashion, a natural vehicle for the controlling theme. Shakespeare makes a much more extensive use of fashion imagery here than in any other play.[1] What is more interesting is the deliberate emphasis given this imagery: all the uses may be regarded as masking other meanings. Some of these images may arise from a fortuitous association of ideas on the author's part, particularly when they occur in isolated sentences, as when Beatrice says that Benedick "wears his faith but as the fashion of his hat; it ever changes with the next block" (i. i. 75–77). But the main uses are integrated with the action itself. Thus one of the marks of Claudio's newly discovered love for Hero is his interest in fashions (ii. iii. 15–19). Benedick is likewise betrayed as a lover by his sudden fancy for strange disguises:

> *Pedro.* There is no appearance of fancy in him, unless it be a fancy that he hath to strange disguises; as to be a Dutchman today, a Frenchman to-morrow; or in the shape of two countries at once, as a German from the waist downward, all slops, and a Spaniard from the hip upward, no doublet. Unless he have a fancy to this foolery, as it appears he hath, he is no fool for fancy, as you would have it appear he is.   (iii. ii. 31–39)

Obviously both of these uses should be reflected in the costuming of the actors. Fashion is not merely a cluster of images here; with its associated images of the cut beard, washed face, perfume, and paint (ll. 42–59), it is the substance of the main part of the scene, occupying the minds of Leonato, Don Pedro, and Claudio: the gentlemen conclude from the outward signs of Benedick's appearance and behavior

---

From "*The Thief 'Deformed' and Much Ado About 'Noting'*" by *James A. S. McPeek.* From Boston University Studies in English, *IV* (*Boston: Department of English, Graduate School, Boston University, 1960*), *68–70. Copyright © 1960 by Boston University. Reprinted by permission of the editor and author.*

[1] Miss Spurgeon finds that images of sporting are more frequent in this play than in any other (*Shakespeare's Imagery,* p. 264). But these images are not so numerous nor so fully developed as those of fashion. Bartlett records sixteen direct references to fashion and fashion-monging in *Much Ado,* over twice as many as in any other play. These references do not include several associated images dealing with clothing and appearance in general. Several of the sporting images, notably those of netting, trapping, and angling, also illustrate the theme of illusion.

that he is in love. In the next instant the complacent vision of Claudio and Don Pedro is to be tested by the deceptions of Don John.

Just as the gentlemen have a scene devoted to the imagery of fashion and its meaning, the ladies also have their scene, in which, though the controlling theme of the play is less overt, it is still suggestively present: the "fine, quaint, graceful, and excellent fashion" of Hero's gown (III. iv. 13–23), worth ten of any glittering show such as that of the Duchess of Milan's gown, may perhaps be taken to represent the reality of Hero herself. Beatrice herself appears in a new guise in this scene. Does Shakespeare mean to suggest further that Beatrice's new fashion, that of her quaintly dissembled love for Benedick, is worth ten of her glittering earlier guise as my Lady Disdain?

In the light of this emphasis on fashion in the imagery, it is interesting that the word *fashion* itself is employed in key incidents as a verb meaning to *shape* or *contrive* events (II. i. 384; II. ii. 47; IV. i. 236). And it is in this sense of contriving or shaping the appearance of things that the most significant use of fashion occurs in this play. What was a set of images suddenly takes shape as a symbol, a personification of the theme, a creature who may be regarded as the shaper and contriver of the tangled web of appearances that composes the fabric of this play.

James Smith

Dogberry and his fellows, of from time to time the victims of syllables like Mrs. Malaprop, are more frequently and more significantly, like the second Mrs. Quickly, the victims of ideas. When Verges speaks of "suffering salvation body and soul," and Dogberry of being "condemned into everlasting redemption," it is impossible they are being deceived merely by similitude of sounds. Rather, they are being confounded by ideas with which, though unfitted to do so, they feel it incumbent upon themselves to cope. Such utterances are of a piece with Dogberry's method of counting; with his preposterous examination of Conrade and Borachio, in which condemnation precedes questioning; with his farewell of Leonato, to whom, in an endeavour to conserve both their dignities, he "humbly gives leave to depart"; with his desire "to be written down an ass," in which the same sense of his own dignity is in conflict with, among other things, a sense that it

From "Much Ado About Nothing: *Notes from a Book in Preparation" by James Smith. From* Scrutiny, *XIII (London: Reprinted by Cambridge University Press, 1963), pp. 243–44. Copyright © 1963 by the syndics of Cambridge University Press. Reprinted by permission of the publisher.*

needs vindication. It is not Mrs. Malaprop, but rather Bottom, who comes to mind here: Bottom who, like Dogberry, is torn between conflicting impulses—whether those of producing his interlude in as splendid a manner as possible, while at the same time showing as much deference as possible to the ladies; or of claiming as his own the "most rare vision" which, as a vision, certainly had been his, while for its rarity it seemed such as could not rightly belong to any man.

In thus addressing themselves to intellectual or moral feats of which they are not capable, Bottom, Mrs. Quickly and Dogberry do of course display a form of pride. Given his attitude towards Verges:

> "a good old man, sir, hee will be talking as they say, when the age is in, the wit is out, God helpe us, it is a world to see. . . ."

Dogberry's pride needs no stressing. It is however no longer a foolish pride; or if foolish, then not with the folly of Mrs. Malaprop, but rather of all the protagonists of drama, comic or tragic, who measure themselves against tasks which ultimately prove too much for them. Perhaps with justice it is to be classified as a form of *hybris,* a comic *hybris;* and if so, then some kind of essential relation between the Dogberry scenes and the tragically inclined scenes of the main plot is immediately suggested.

The suggestion is strengthened, once Dogberry's strength rather than his weakness, his triumphs rather than his failures, are considered. For he has established himself as Constable of Messina, not only to the content of his subordinates, but with the tolerance of his superiors. In this respect he is no longer to be compared with Bottom—who, it is to be feared, would never gain a firm footing, however humble, at the court of Theseus—but with Falstaff, a character of greater importance. Unlike Bottom, Dogberry and his companions have taken fairly accurate measure both of themselves and of those who surround them; so that, if swayed by *hybris* in a certain degree, they take care that this degree shall fall sort of destructive. For example, they are quite clear "what belongs to a Watch": they will "sleep rather than talk"; rather than bid a man stand against his will, they will let him go and thank God they are rid of a knave; rather than take a thief, they will "let him shew himselfe for what he is," and steal out of their company. In short, they will exert themselves, or fight, no longer than they see reason: to adapt Poins's words. Indeed, in this matter they are more consistent than Falstaff, who, in dismissing Prince Henry as "a Fellow, that never had the Ache in his shoulders," is for once allowing himself to be puffed up by *hybris.* In his boasts to Shallow, Falstaff betrays not a little of a Bottom-like recklessness:

Master *Robert Shallow,* choose what Office thou wilt in the Land, 'tis thine . . . Boote, boote, Master *Shallow,* I know the young King is sick for mee. . . .

And discomfiture of course follows. Whereas Dogberry has perfectly accommodated himself to those on whom he depends, making their ideals his own. His list of qualifications is revealing:

I am a wise fellow, and which is more, an officer, a householder, and which is more, as pretty a peece of flesh as any in Messina, and one that knowes the Law, goe to, and a rich fellow enough, goe to, and a fellow that hath had losses, and that hath two gownes, and everything handsome about him.

It needs little acquaintance with the Leonato circle to realize that for them too it is a principal concern that everything, as far as possible, shall remain "handsome about them."

Walter N. King

It is here [in the church scene] that the social abnormality of aristocratic society in Messina is exposed once and for all for what it is— shallow and perverse application of a standard of behavior that is both automatic and uncharitable. In part, critical misunderstanding of this scene has sprung from failure to realize that the deception by Don John and Borachio of Claudio and Don Pedro into the belief that Hero is sexually loose is symbolic as well as psychological. Inability to see clearly at night is a common human trait, but in Claudio and Don Pedro it symbolizes the dominant trait of aristocratic folk in Messina, in whom failure of physical eyesight is equivalent to moral confusion. Those who marry according to the philosophy of *caveat emptor,* like Claudio, are bound to be predisposed to sexual distrust, while their depreciation of love and marriage to the level of the market-place inevitably leads them to believe in virginity as the principal attribute of a bride-to-be.

Claudio's determination to expose Hero in church is quite in line with the social usage of his society, which accepted as legitimate harsh reprisal for sexual fraud, but he also exposes his general moral blind-

From "Much Ado About Something" *by Walter N. King. From* Shakespeare Quarterly, *XV (New York: The Shakespeare Association of America, Inc., 1964), 150–51. Copyright © 1964 by The Shakespeare Association of America, Inc. Reprinted by permission of the editor.*

ness. And the immediate compliance of Don Pedro (III. ii. 126–130) indicates that Claudio's decision, however lacking in Christian charity, should not be reckoned a complete social abnormality. All those who reject Hero, even Leonato, assume they are justified, and they all behave melodramatically, just as shallow human beings are always inclined to thunder for justice in a social crisis when wounded pride, far more than moral shock, begins to stem up their ethical consciousness.

Nevertheless, Claudio's self-righteousness exposes a serious flaw in the social code: the superficiality of a value system that mistakes sexual purity for love is shown up in all its heartless folly. At the same time, the concurrent movement away from superficiality in Beatrice and Benedick, already under way, suggests how witlessness can be exchanged for wisdom. Stupidity versus intelligence is the underlying theme of the church scene and is dramatized by means of a typical Shakespearian problem in epistemology: under what conditions can the senses be trusted to provide accurate data for substantive knowledge of human character? To what degree do objective and subjective ways of knowing lead to rock-bottom truth about people we think we are familiar with?

The dialectic begins in Claudio's ironic reflection upon human presumption: "O, what men dare do! what men may do, what men daily do, not knowing what they do!" (IV. i. 19–21). His folly—tragedy to his social peers—is to confuse what appears to his eyes, Hero's external look of innocence, with what appears to his mind, her alleged promiscuity. "Would you not swear, All you that see her, that she were a maid, By these exterior shows?" (IV. i. 29–41). The either/or mentality of the mediocre mind trying to think erupts in a burst of hackneyed metaphor:

> You seem to me as Dian in her orb,
> As chaste as is the bud ere it be blown,
> But you are more intemperate in your blood
> Than Venus, or those pamp'red animals
> That rage in savage sensuality.   (IV. i. 58–62)

Some lines later comes a saving note of doubt: "Are our eyes our own?" (IV. i. 72). Claudio is on the verge of learning the first lesson of the Platonic theory of knowledge, that the senses may deceive. (His early confession to Benedick that Hero "is the sweetest lady that ever I looked on" [I. i. 189], has now been transformed into the false assumption that her "blush is guiltiness, not modesty" [IV. i. 43].) But he is far from grasping the second lesson, that the senses are sometimes trustworthy. Appearance can be reality.[1] As a consequence, he leaps to a false conclusion about Hero, owing to a confusion of mind that springs naturally enough from reliance upon second-rate values.

But Claudio is no worse than those who, knowing Hero better than he, take at face value the "fact" of her depravity. In twenty-three impassioned lines dripping with the sentimentality and bombast an unexamined moral code can produce, Leonato sermonizes on the theme: "Why ever wast thou lovely in my eyes?" (IV. i. 121–144). "Let her die," he urges (IV. i. 155), and insists, "She not denies it" (IV. i. 174), in the face of Hero's flat declaration to Claudio, "I talked with no man at that hour, my lord" (IV. i. 87). Leonato's allegiance to a dessicated social norm continues even after Friar Francis outlines a means for retrieving Hero's reputation. As hyperbolic as Claudio, Leonato also illustrates the truth of Beatrice's summary estimate of the male world of Messina: "But manhood is melted into courtesies, valor into compliments, and men are turned into tongues, and trim ones too. He is now as valiant as Hercules that only tells a lie, and swears it" (IV. i. 20–24). No longer fooled by words, she longs to be a man in a society in which the traditional concept of manhood has become debased.

She, too, along with Benedick, contributes to the dialectic. Whereas Beatrice *knows* instinctively that Hero "is belied" (IV. i. 147), Benedick's reaction is "I know not what to say" (IV. i. 146), a way to begin to know. His earlier brag, "I can see yet without spectacles" (I. i. 191), has ceased to be an immodest claim, now that his faith in verbal gymnastics has vanished. His is the first sensible question to be asked, "Lady [Beatrice], were you her bedfellow last night?" (IV. i. 148)—a way of knowing through research; and only he is keen-eyed enough to suspect Don John's complicity in the slander—a way of knowing through hypothesizing. Together with the behavior of Beatrice and Friar Francis, whose reasoned faith in Hero's innocence is grounded in objective observation combined with extensive experience of human nature (another way of knowing), Benedick's behavior diverges sharply from the inadequate norms of Messina toward a revitalization of the norms that will culminate in Hero's restoration.

Such revitalization is difficult, demanding as it does the development of insight in people accustomed to see dimly. Those who can be tricked into seeing what is not obvious (Hero's "guilt") must be tricked into

---

[1] Kerby Neill ["More Ado About Claudio: An Acquittal for the Slandered Groom," *Shakespeare Quarterly*, III (1952)], p. 93, makes a somewhat similar comment, but in terms of the traditional conflict between reason and emotion. I find it hard, however, to accept Neill's description of Claudio as a somewhat idealistic, if naive, young man.

seeing what is plain (her innocence); hence, Friar Francis' plan, based upon the psychological fact that superficial people have only a limited capacity for change, to reform Claudio's vision (and so his thinking) by deceiving him into the belief that Hero is dead.

John Palmer

"Much Ado About Nothing," as already suggested, is Shakespeare's nearest approach to the comedy of manners: the wit-combat between its predestinate lovers, besides being a favourite device of the author for establishing a lively familiarity between the parties, commits him to what is, in effect, the stock situation in a type of play which Congreve brought to perfection a century later. Beatrice and Benedick bickering their way into matrimony, are quite obviously assuming for our pleasure a social attitude which we know to be at variance with their true feelings and with the destiny which awaits them at the close.

In the pure comedy of manners as practised by Congreve sentiment or passion is conveyed by means of elegant understatement or downright contradiction:

> *Mrs. Millamant.* I won't be called names after I'm married; positively I won't be called names.
> *Mirabel.* Names!
> *Mrs. Millamant.* Ay, as wife, spouse, my dear, joy, jewel, love, sweetheart, and the rest of that nauseous cant in which men and their wives are so fulsomely familiar—I shall never bear that. Good Mirabel, don't let us be familiar or fond, nor kiss before folks, like my lady Fadler and Sir Francis, nor go to Hyde Park together the first Sunday in a new chariot, to provoke eyes and whispers, and then never to be seen there together again; as if we were proud of one another the first week, and ashamed of one another ever after. Let us never visit together nor go to a play together; but let us be very strange and well-bred: let us be as strange as if we had been married a great while; and as well-bred as if we were not married at all.

Millamant and Mirabel affect to have outgrown and overcome the promptings of nature.

The real point of the joke is that man is pretending to be civilised. This is the stock situation of the comedy of manners. The elaborate

From "Beatrice and Benedick, Much Ado About Nothing" by John Palmer. From Comic Characters of Shakespeare (*London: Macmillan & Co. Ltd., 1946*), *pp. 117–19, 120–21. Copyright © 1946 by Macmillan & Co. Ltd. Reprinted by permission of the publisher.*

ritual of society is a mask through which the natural man is comically seen to look. The comedy of Millamant is that she is about to be married as a woman, and that she talks of her marriage merely like a person in society. In the comedy of manners men and women are seen holding reality away, or letting it appear only as an unruffled thing of attitudes. Life is here made up of exquisite demeanour. Its comedy grows from the incongruity of human passion with its cool, dispassionate and studied expression. It ripples forth in ironic contemplation of people born to passion high and low, posing in the social mirror. This is the real justification of the term "artificial comedy" as applied to the plays of Congreve. We are born naked into nature. In the comedies of Congreve we are born again into civilisation and clothes. We are no longer men; we are wits and a peruke. We are no longer women; we are ladies of the tea-table. Life is absurdly mocked as a series of pretty attitudes and sayings. Hate is absurdly mirrored in agreeably bitter scandal. Perplexity and wonder are seen distorted in the mechanical turns of a swift and complicated plot. Always the fun lies in a sharp contrast between man civilised and the genial primitive creature peeping through.

*   *   *

Congreve wrote the undiluted comedy of manners, distilling pure water from the living spring. His characters must sustain to the end their manifest pretences that they have no feeling deeper than an epigram may carry; no aspiration higher than a fine coat may express; no impulse stronger than a smile may cover; no joy more thrilling than a nod may contain; no sorrow deeper than a pretty oath may convey. Shakespeare's comedy, on the other hand, consists in elaborating these pretences in order that they may at the right moment be effectively exploded. Beatrice and Benedick, who begin by seeming least likely of any in Messina to betray a genuine emotion, must in the end uncover their hearts.

Northrop Frye

Benedick and Beatrice in *Much Ado* are . . . mechanical comic humors, prisoners of their own wit, until a benevolent practical joke enables their real feelings to break free of their verbal straitjackets.

From "The Triumph of Time" by Northrop Frye. From A Natural Perspective: The Development of Shakespearean Comedy and Romance *(New York: Columbia University Press, 1965), p. 81. Copyright © 1965 by Columbia University Press. Reprinted by permission of the publisher.*

This benevolent practical joke is in contrast to the malevolent one
that Don John plays on Claudio, which, though far more painful in
its effects, operates according to the same comic laws. Claudio becomes
engaged to Hero without also engaging his loyalty: he retains the desire to be rid of her if there should be any inconvenience in the arrangement, and this desire acts precisely like a humor, blinding him
to the obvious facts of his situation. In his second marriage ceremony
he pledges his loyalty first, before he has seen the bride, and this re-
leases him from his humorous bondage.

Wylie Sypher

Shakespeare's plays, says Meredith, are saturated with the golden
light of comedy—the comedy that is redemptive as tragedy cannot
be. Consider what happens in *Much Ado About Nothing* when Benedick makes the startling comic discovery that he himself, together
with the other mistaken people in the play, is a fool. Here is a moral
perception that competes with tragic "recognition." The irony of
Benedick's "recognition" is searching, for he has boasted, all along,
that he cannot find it in his heart to love any of Eve's daughters,
least of all Beatrice. And Beatrice, for her part, has avowed she will
never be fitted with a husband until God makes men of some other
metal than earth. Both these characters are too deep of draught to
sail in the shoal waters of sentimentality, and both have bravely laid
a course of their own far outside the matchmaking that goes easily
on in Messina. Each is a mocker, or eiron; but in being so, each
becomes the boaster (alazon) betrayed into the valiant pose that they
are exempt from love. Then they both walk, wide-eyed, like "proud"
Oedipus, into the trap they have laid for themselves. There they see
themselves as they are. When Benedick hears himself called hard-
hearted he suffers the bewilderment of comic discovery and knows
that his pose as mocker is no longer tenable. So he turns his scornful
eye inward upon his own vanity: if Beatrice is sick for love of his
ribald self he must give up his misogyny and get him a wife. He
yields himself, absurdly, to Beatrice, saying "Happy are they that
hear their detractions and can put them to mending." At the extreme
of his own shame Benedick is compelled to see himself as he sees
others, together along a low horizon. Thus occur the comic purgation,

From "The Social Meanings of Comedy" and "Appendix" by Wylie Sypher. From
The Meanings of Comedy (*New York: Doubleday and Company, 1956*), *pp. 253–54.*
[*This is a distillation of the author's "Nietzsche and Socrates in Messina,"* Partisan
Review, *XVI (1949), 702–13.*] *Reprinted by permission of the author.*

the comic resignation to the human lot, the comic humbling of the proud, the comic ennobling after an act of blindness. Those who play a comic role, like Benedick or Berowne or Meredith's Sir Willoughby Patterne, wrongheadedly are liable to achieve their own defeat and afterwards must hide their scars. The comic and the tragic heroes alike "learn through suffering," albeit suffering in comedy takes the form of humiliation, disappointment, or chagrin, instead of death.

Chronology of Important Dates

<table>
<tr><td></td><td align="center">Shakespeare</td><td align="center">The Age</td></tr>
<tr><td>1558</td><td></td><td>Accession of Elizabeth I.</td></tr>
<tr><td>1564</td><td>April 26: William Shakespeare christened at Stratford-Upon-Avon.</td><td></td></tr>
<tr><td>1572</td><td></td><td>John Donne born.</td></tr>
<tr><td>1576–77</td><td></td><td>"The Theatre" and "The Curtain" opened.</td></tr>
<tr><td>1582</td><td>November 27: Marriage with Anne Hathaway licensed.</td><td></td></tr>
<tr><td>1583</td><td>May 26: Christening of daughter Susanna.</td><td></td></tr>
<tr><td>1585</td><td>February 2: Hamlet and Judith Shakespeare christened.</td><td></td></tr>
<tr><td>1586</td><td></td><td>Sir Philip Sidney dies.</td></tr>
<tr><td>1588</td><td></td><td>Defeat of the Spanish Armada.</td></tr>
<tr><td>1588–94</td><td>The Comedy of Errors, Love's Labor's Lost, the first tetralogy of history plays, The Taming of the Shrew, and Two Gentlemen of Verona performed.</td><td>George Peele's The Old Wives' Tale, John Lyly's Endimion and Mother Bombie, Robert Green's Friar Bacon and Friar Bungay performed.</td></tr>
<tr><td>1590</td><td></td><td>Edmund Spenser's The Faerie Queene, Books I–III, published.</td></tr>
<tr><td>1593</td><td></td><td>Christopher Marlowe's Doctor Faustus performed; Marlowe dies.</td></tr>
</table>

1594	Becomes a shareholder in The Lord Chamberlain's Company.	
1594–97	*Romeo and Juliet, Richard II, King John, A Midsummer Night's Dream,* and *The Merchant of Venice* performed.	
1596	Hamnet dies; Shakespeare's father granted arms.	Books IV–VI of *The Faerie Queene* published.
1597	May 4: Purchases New Place, Stratford.	Francis Bacon's *Essays* published.
1597–98	*Henry IV,* Parts I and II, performed.	
1598		Ben Jonson's *Every Man in His Humor,* in which Shakespeare acts, performed.
1598–1600	*Much Ado About Nothing* performed.	
1599		"The Globe" built.
1599–1600	*Henry V, Julius Caesar, As You Like It,* and *Twelfth Night* performed.	
1601	September 8: Father buried.	
1601–4	*Hamlet, The Merry Wives of Windsor, Troilus and Cressida, All's Well That Ends Well,* and *Othello* performed.	
1603		Elizabeth dies; accession of James I.
1604–9	*Measure for Measure, King Lear, Macbeth, Antony and Cleopatra, Timon of Athens,* and *Coriolanus* performed.	
1605		The Gunpowder Plot.
1606		Ben Jonson's *Volpone* performed.
1608–13	*Pericles, Cymbeline, A Winter's Tale,* and *The Tempest* performed.	

1616   February 10: Judith marries
            Thomas Quiney. March 25:
            Makes his will. April 23: Dies
            at Stratford.

1623   First Folio published.

Notes on the Editor and Contributors

WALTER R. DAVIS, the editor, is Professor of English at The University of Notre Dame. Among his publications are a study of Sir Philip Sidney's *Arcadia* (1965) and an edition of Thomas Campion (1967); he has recently completed a study of Elizabethan fiction.

W. H. AUDEN, whose latest volume of verse is *About the House* (1965), has among his volumes of criticism *The Enchafed Flood* (1950) and *The Dyer's Hand* (1962). In collaboration with Chester Kallman, he has translated the libretto of Mozart's *The Magic Flute* (1956) and composed the libretto for Igor Stravinsky's opera *The Rake's Progress* (1951). His achievement is assessed in a volume in the Twentieth Century Views series.

JOHN CRICK, Lecturer in English and American Studies at Didsbury College of Education, Manchester, England, has published poetry and literary criticism in various journals and has broadcast over the BBC. He is currently working on a book on Alexander Pope.

FRANCIS FERGUSSON, who is widely recognized as the most influential modern critic of the drama, is Professor of Comparative Literature at Rutgers University. His books include *The Idea of a Theater* (1949), *Dante's Drama of the Mind* (1952), *The Human Image in Dramatic Literature* (1957), and *Poems* (1962).

NORTHROP FRYE is University Professor of English at the University of Toronto. He is author of *Fearful Symmetry: A Study of William Blake* (1947), *Anatomy of Criticism* (1957), *The Well-Tempered Critic* (1963), *Fables of Identity: Studies in Poetic Mythology* (1963), *The Educated Imagination* (1964), *The Return of Eden: Five Essays on Milton's Epics* (1965), and *A Natural Perspective* (1965).

HAROLD C. GODDARD was Head of the Department of English at Swarthmore College from 1909 to 1946. His book *The Meaning of Shakespeare* was published posthumously in 1951.

DOROTHY C. HOCKEY, Associate Professor of English at Lake Erie College, has contributed several articles to *Shakespeare Quarterly*.

DAVID HOROWITZ is the author of *Shakespeare: An Existential View* (1965) and, in a totally different vein, *The Free World Colossus: A Critique of American Foreign Policy in the Cold War* (1965).

ROBERT GRAMS HUNTER is Associate Professor of English at Dartmouth College.

WALTER N. KING, Associate Professor of English at The University of Montana, has published articles on Shakespeare and John Lyly.

WILLIAM G. McCOLLOM, Professor of English at Western Reserve University, has directed several plays, written essays on Shakespeare and George Chapman, and published a book, *Tragedy* (1957).

JAMES A. S. McPEEK, Professor of English at The University of Connecticut, has written, besides several articles, *Catullus in Strange and Distant Britain* (1939) and edited, with Robert C. Baldwin, *An Introduction to Philosophy Through Literature* (1950).

JOHN PALMER had published forty detective and historical novels in collaboration with Hilary St. George Sanders (under the shared pseudonym "Francis Beeding") by his death in 1944. He was author in his own name of many books on drama, including *The Comedy of Manners* (1913), *George Bernard Shaw: Harlequin or Patriot?* (1915), *Political Characters of Shakespeare* (1945), and *Comic Characters of Shakespeare* (1946), the last two being parts of a projected trilogy cut off by his death.

A. P. ROSSITER, who died in 1957, was Lecturer at Durham University and Cambridge University; he also lectured frequently at the Shakespeare Summer School at Stratford and broadcast over the BBC. He was author of a series of articles on Shakespeare in the *Durham University Journal*, *English Drama from Early Times to the Elizabethans* (1950), and *Angel with Horns* (1961), a posthumous collection of lectures.

JAMES SMITH is best known for a series of articles on various subjects, including the French *symboliste* poets and English "metaphysical" poetry as well as Elizabethan drama, which he contributed to the English journal *Scrutiny* from 1933 to 1946.

GRAHAM STOREY, besides lecturing on Shakespeare, has edited the letters of Charles Dickens in collaboration with Madeline House (1965) and written a history of the Reuters news service (1951).

WYLIE SYPHER is Chairman of the Department of English at Simmons College. Among his many books are *Four Stages of Renaissance Style* (1955), *Comedy* (1956), *Rococo to Cubism in Art and Literature* (1960), and *Loss of the Self* (1962).

VIRGIL THOMSON, composer and music critic, had composed chamber music, symphonies, songs, oratorios, and sound tracks for motion pictures; the best known of his works is the opera *Four Saints in Three Acts,* with libretto by Gertrude Stein (1934). Among his books are *The Musical Scene* (1945), *The Art of Judging Music* (1948), and *Music Left and Right* (1951).

JAMES J. WEY is Assistant Professor of English at The University of Detroit.

Selected Bibliography

It should be noted that, in addition to the following studies, the essays from which excerpts were taken for "View Points" will repay reading in their entirety.

Barber, C. L., *Shakespeare's Festive Comedy: A Study of Dramatic Form and Its Relation to Social Custom.* Princeton: Princeton University Press, 1959. The most illuminating recent book on the comedies, stressing the ways in which their roots in May-day rituals and the like influence their structure and total effect; while Barber does not treat *Much Ado* directly, he asserts (p. 222) that much of what he writes about *Twelfth Night* and other comedies applies to it as well.

Bradbrook, M. C., *Shakespeare and Elizabethan Poetry.* London: Chatto & Windus, 1951. The discussion of *Much Ado* centers on the theme of self-deception and on the intuitive powers of love.

Brown, John Russell, *Shakespeare and His Comedies.* London: Methuen, 1957. Chapter IV discusses the difficulty of finding certainty in love in *Much Ado* and other comedies.

Craik, T. W., "*Much Ado About Nothing*," *Scrutiny*, XIX (1953), 297–316. A careful scene-by-scene analysis showing the play's unity to reside in the theme of error.

Evans, Bertrand, *Shakespeare's Comedies.* London: Oxford University Press, 1960. Chapter III treats the presentation of man's propensity for misunderstanding in *Much Ado, As You Like It,* and *The Merry Wives of Windsor.*

Everett, Barbara, "*Much Ado About Nothing*," *The Critical Quarterly*, III (1961), 319–35. A discussion of the war of the sexes in *Much Ado.*

Neill, Kerby, "More Ado About Claudio: An Acquittal for the Slandered Groom," *Shakespeare Quarterly*, III (1952), 91–107. Neill grounds his defense both on the context of the play, which, he insists, does not condemn Claudio, and on changes Shakespeare made in his sources in order to make Claudio a more appealing figure.

Phialas, Peter G., *Shakespeare's Romantic Comedies: The Development of Their Form and Meaning.* Chapel Hill: The University of North Carolina Press, 1966. Chapter VII constitutes an extensive general commentary on *Much Ado.*

Prouty, Charles T., *The Sources of "Much Ado About Nothing."* New Haven, Conn.: Yale University Press, 1950. This is not only an exhaustive study of the sources, but also a penetrating commentary on the play in the light of Shakespeare's alterations of his sources.

Stevenson, David Lloyd, *The Love-Game Comedy.* New York: Columbia University Press, 1946. A study of the traditional conflict between the theory and the reality of love as presented in *Much Ado* and the other romantic comedies. Chapter XII contrasts *Much Ado* with *Troilus and Cressida* as comic and satiric views of courtship, respectively.

Wilson, J. Dover, *Shakespeare's Happy Comedies.* Evanston: Northwestern University Press, 1962. Contains a lengthy discussion of *Much Ado* centering on the "hide-and-seek" action in the play.

TWENTIETH CENTURY
INTERPRETATIONS

MAYNARD MACK, *Series Editor*
Yale University

NOW AVAILABLE
Collections of Critical Essays
ON

ADVENTURES OF HUCKLEBERRY FINN
ALL FOR LOVE
THE AMBASSADORS
ARROWSMITH
AS YOU LIKE IT
BLEAK HOUSE
THE BOOK OF JOB
THE CASTLE
DOCTOR FAUSTUS
DUBLINERS
THE DUCHESS OF MALFI
EURIPEDES' ALCESTIS
THE FROGS
GRAY'S ELEGY
THE GREAT GATSBY
GULLIVER'S TRAVELS
HAMLET
HARD TIMES
HENRY IV, PART TWO
HENRY V
THE ICEMAN COMETH
JULIUS CAESAR
KEATS'S ODES

(continued on next page)

(*continued from previous page*)

Lord Jim

Much Ado About Nothing

Oedipus Rex

The Old Man and the Sea

Pamela

The Playboy of the Western World

The Portrait of a Lady

A Portrait of the Artist as a Young Man

The Rime of the Ancient Mariner

Robinson Crusoe

Samson Agonistes

The Scarlet Letter

Sir Gawain and the Green Knight

The Sound and the Fury

The Tempest

Tom Jones

Twelfth Night

Utopia

Walden

The Waste Land

Wuthering Heights